Kulturelle Figurationen: Artefakte, Praktiken, Fiktionen

Kultur gilt—neben Kategorien wie Gesellschaft, Politik, Ökonomie—als eine grundlegende Ressource sozialer Semantiken, Praktiken und Lebenswelten. Die Kulturanalyse ist herausgefordert, kulturelle Figurationen als ebenso flüchtige wie hegemoniale, dynamische wie heterogene, globale wie lokale und heterotope Phänomene zu untersuchen. Kulturelle Figurationen sind Produkt menschlichen Zusammenlebens und bilden zugleich die sinnstiftende Folie, vor der Vergesellschaftung und Institutionenbildung stattfinden. In Gestalt von Artefakten, Praktiken und Fiktionen sind sie uneinheitlich, widersprüchlich im Wortsinn und können doch selbst zum sozialen Akteur werden. Die Reihe »Kulturelle Figurationen: Artefakte, Praktiken, Fiktionen« untersucht kulturelle Phänomene in den Bedingungen ihrer Produktion und Genese aus einer interdisziplinären Perspektive und folgt dabei der Verflechtung von Sinnzusammenhängen und Praxisformen. Kulturelle Figurationen werden nicht isoliert betrachtet, sondern in ihren gesellschaftlichen Situierungen, ihren produktionsästhetischen und politischen Implikationen analysiert. Die Reihe publiziert Monographien, Sammelbände, Überblickswerke sowie Übersetzungen internationaler Studien.

Aïda C. Terblanché-Greeff · Elisabeth Alm ·
Jörn Ahrens
Editors

Climate Change and the Postcolonial

Springer VS

Editors
Aïda C. Terblanché-Greeff
North-West University
Potchefstroom, South Africa

Elisabeth Alm
University of Freiburg
Freiburg, Germany

Jörn Ahrens
Justus Liebig University
Giessen, Germany

ISSN 2567-4242 ISSN 2625-0896 (electronic)
Kulturelle Figurationen: Artefakte, Praktiken, Fiktionen
ISBN 978-3-658-49327-1 ISBN 978-3-658-49328-8 (eBook)
https://doi.org/10.1007/978-3-658-49328-8

Bildzeichen: Falk Nordmann, Berlin

This Springer VS imprint is published by the registered company Springer Fachmedien Wiesbaden GmbH, part of Springer Nature.
The registered company address is: Abraham-Lincoln-Str. 46, 65189 Wiesbaden, Germany

Contents

Editors and Contributors

Jörn Ahrens is a Professor of Cultural Sociology with Focus on the Transformation of Culture at Justus Liebig University, Giessen, Germany, and Extraordinary Professor of Social Anthropology at North-West University, South Africa. His primary research areas include sustainability, nature, and the Global South, as well as popular culture (film and comics), society and violence, and critique of modernity. He is the editor of the book '*Climate Change Epistemologies in Southern Africa: Social and Cultural Dimensions*' (London & New York: Routledge. 2023. https://doi.org/10.4324/9781003180814), with Ernst Halbmayer. Other publications include '*Gone are the Days: The Recent Western, the Imagination of Violence, and the Abandoning of the Hero*' (in: M. von Koppenfels and M. Mühlbacher [eds.]. Adventure at Arms. On the Narrative Formation of Violence, Paderborn: Brill/Fink, 2024. https://doi.org/10.30965/9783846768709_014), and '*On the History of Fascination with Violence: Mediations of the Immediate*' (in: J. Ecarius and J. Bilstein [eds.]: Violence—Reason—Fear: Interdisciplinary Approaches and Theoretical Approaches, Wiesbaden: Springer VS, 2023. https://doi.org/10.1007/978-3-658-40886-2_6). He is also a member of the EUPeace Climate Change Research Hub at the universities of Gießen and Marburg, Germany.

Elisabeth Alm is an MA graduate in Global Studies at the University of Freiburg (Germany), Facultad Latinoamericana de Ciencias Sociales (Argentina), and Chulalongkorn University (Thailand). She received a BA in Social Sciences from Justus Liebig University, Germany. From 2022 to 2024, she served as a research assistant on the project *Social Climate Change Impacts and Sustainability Innovation in Southern Africa and Northern South America* (NISANSA) at Justus Liebig University of Giessen and Philipps University of Marburg. In cooperation with the University of Botswana and North-West University (South Africa), the project researched climate knowledge as a cultural practice and its translation into the politics of transformation. Informed by her field research in various sub-Saharan countries, her research interests include the impacts of social climate change, decoloniality, and global power structures.

Yolandi M. Coetser is an Associate Professor and director in the School of Philosophy at North-West University, South Africa. Her primary research focus is on environmental philosophy, with a special focus on animal ethics and conservation philosophy. Some of her notable publications include '*An African Ethical Perspective on South Africa's Regulatory Frameworks Governing Animals in Research*' (Studies in History and Philosophy of Science, 2022. https://doi.org/10.1016/j.shpsa.2022.01.021), '*Exploring the Legality of Artists' Use of Animals: Ethical Considerations and Legal Implications*' (Arts, 2023. https://doi.org/10.3390/arts12040155), and '*No Monkeys in My Circus: Philosophical Perspectives Towards a Legal Ban of Circus Animals in South Africa*' (Animal Law-Animal Legal and Policy Studies, 2023. https://doi.org/10.36151/DALPS.006). She is also the founder and chair of the *Southern African Society of Environmental Philosophy.*

Gabriel Faimau is a Professor of Sociology in the Department of Sociology at the University of Botswana. He is also a Visiting Associate Professor at the Botswana International University of Science and Technology (BIUST), and an Extraordinary Researcher in the Unit for Reformational Theology and the Development of the South African Society at North-West University in South Africa. His main research interests include climate change communication, digital society and culture, new media, and religious practice, supplemented by a focus on sustainability issues in Botswana. His recently co-authored various publications, which include '*COVID-19 and the Digitalisation of Food Supply and Delivery: A Case of Botswana*' with M.K. Phuthegelo (in: L.E. Espeter and L. Hering [eds.], Commodity Chains Under Pressure: Resilience through Coping, Adaptation, and Transformation, Cheltenham: Edward Elgar Publishing, 2025. https://doi.org/10.4337/9781035355303.00015) and '*Print Media Coverage and the Socio-Contextual Representation of Climate Change in Botswana*' with N. Sello and E. Nkhukhu-Orlando (in: J. Ahrens and E. Halbmayer [eds.], Epistemologies of Climate Change, London: Routledge, 2023. https://doi.org/10.4324/9781003180814-9). He is also head of SMUS Botswana as one of the lead partners of the *Global Center of Spatial Methods for Urban Sustainability* (GCSMUS).

Michelle Pressend is the academic coordinator for the African Regional Hub of the *Transnational Centre for Just Transitions in Energy, Climate, and Sustainability* (TRAJECTS), a Just Transitions initiative. This project operates under the Energy Systems Research Group at the University of Cape Town (UCT, South Africa), and is a partnership with two universities in Colombia and Germany. She lectured in Environmental Sociology at UCT's Sociology Department. Before embarking on her academic career, she dedicated over twenty years to research, policy analysis, and activism focused on environmental and socioeconomic justice, primarily in the non-governmental sector. Her research interests centre on decoloniality and de-centring Western modernity's extractive values in energy transitions and responses to climate change. Her publications include '*Green Masquerade: Neo-liberalism, Extractive Renewable Energy Transitions, and the "Good" Anthropocene in South Africa*' (International Development Policy, 2023. https://doi.org/10.4000/pol-dev.5695) and '*The Fourth Industrial Revolution and Energy Transitions: Channelling Renewable Energy into New Frontiers of Capital Accumulation*' (in: A. Patel, P. Mthembu, and M. Hlatshwayo [eds.]. UNISA Press, 2023. https://doi.org/10.1080/00358533.2024.2438534).

Johann W. N. Tempelhoff is an Extraordinary Professor at North-West University's Focus Area on Social Transformation (FAST) in South Africa. He is an environmental historian working in the field of the cultural dynamics of water, in transdisciplinary research modes. Much of his work focuses on issues of water and sanitation problems in both urban and rural areas of several South African provinces. He is author/co-author and/or editor of more than 70 peer reviewed articles, 10 book chapters, 33 research reports, and 16 book publications. More specifically, he authored the book '*South Africa's Water Governance Hydraulic Mission (1912-2008) in a WEF Nexus Context*' (2018, AOSIS, Cape Town. https://doi.org/10.4102/aosis.2018.BK75), and the article '*Emfuleni's Wastewater Crisis, 2018-2021: The History of a Vaal Sub-Catchment Problem*' (New Contree, 2021, 86: 27–54). Besides his ongoing research, which currently focuses on climate change and water, he also provides postgraduate student guidance.

Aïda C. Terblanché-Greeff was previously appointed as an Extraordinary Researcher at the Focus Area on Social Transformation (FAST) and is currently a senior lecturer at the School of Philosophy at North-West University in South Africa. Her research is generally interdisciplinary, and her interests include

African philosophy, social philosophy, sustainability, disaster risk reduction, and ethics. Her publications include '*Ubuntu and Environmental Ethics: The West Can Learn from Africa When Faced with Climate Change*' (in: M. Chemhuru [ed.]. African Environmental Ethics: A Critical Reader. Dordrecht: Springer. 2019. https://doi.org/10.1007/978-3-030-18807-8_7) and '*Integrating Deep Ecology, Degrowth, and Ubuntu: Foundations for an Ecosophy*' (in: K.U. Abudu, K.G. Behrens, and E. Imafidon [eds.]. African Philosophy and Deep Ecology. Routledge, 2026. https://doi.org/10.4324/9781003480914). She also developed the *Metzian Ubuntu Inventory* and the *African Time Inventory* through her empirically engaged work.

Introduction: The Postcolonial Legacy of Climate Change

1

Jörn Ahrens, Aïda C. Terblanché-Greeff, and Elisabeth Alm

In 2013, the United Nations Climate Change (UNCC) division established six Regional Collaboration Centres (RCCs) worldwide. According to the UNCC brochure on RCCs, these centres are primarily meant to 'partner with regional host organizations to mobilize networks and enable climate action' (UN, 2023: 2). A look at the world map quickly reveals that all six centres are located in the so-called Global South as they are intended to 'source know-how and resources to drive clean development in developing countries' (UN, 2023) as 'regional collaboration drives global progress. By connecting governments and stakeholders, RCCs raise national ambition, support regional collaboration and accelerate action towards the Paris Agreement goals' (UN, 2023). One major activity of the RCCs is the organisation of Regional Climate Weeks, which should 'offer a platform to showcase regional action, engage with experts and build partnerships' (UN, 2023). On a regional level, in developing or even emerging countries, the RCC Regional Climate Weeks seem to be well established and indeed of considerable importance for connecting actors in climate change action and establishing the topic of climate

J. Ahrens
Justus Liebig University, Giessen, Germany
e-mail: Joern.Ahrens@sowi.uni-giessen.de

A. C. Terblanché-Greeff (✉)
North-West University, Potchefstroom, South Africa
e-mail: Aida.Botha@nwu.ac.za

E. Alm
University of Freiburg, Freiburg, Germany
e-mail: Elisabeth-alm@gmx.de

A. C. Terblanché-Greeff et al. (eds.), *Climate Change and the Postcolonial*, Kulturelle Figurationen: Artefakte, Praktiken, Fiktionen,
https://doi.org/10.1007/978-3-658-49328-8_1

change in the countries in which they take place. However, the UNCC did not hold Regional Climate Weeks in 2024 due to a shortfall in funding, which has caused discomfort among various stakeholders. One effect of this decision would be 'less inclusion for developing-country voices' as 'the world's most vulnerable' (Marrumbe, 2024) would lose an important forum, which would render 'UN climate talks [. . .] no longer fit for purpose' (Marrumbe, 2024). Speaking of people acting on the frontlines of the climate crisis, Dulce Marrumbe, head of partnerships and advocacy at the WaterAid regional office for Southern Africa, points out that RCCs provide a

> platform for those shouldering the heaviest burden of the climate crisis – such as women and girls, people experiencing marginalisation, and Indigenous communities – to share their experiences, expertise, and unique perspectives. [. . .] With leadership and participation from those vulnerable communities and groups, we are all better equipped to adapt to our changing climate – and to ensure that everyone, everywhere has climate-resilient water, sanitation and hygiene. (Marrumbe, 2024)

This cancellation of an important forum for climate change action in the Global South came to the fore due to 'governments' failure to provide enough money' (Marrumbe, 2024). Naturally, the most important donors are the governments of wealthier nations, primarily located in the so-called Global North. Here, it seems, the funding of Global South actors is not a priority in transnational governance agendas. How decisive such support would be is illustrated by a 2019 survey carried out by the pan-African research network *Afrobarometer*. Here it is made clear that African countries are among those most vulnerable to climate change worldwide (Selormey and Logan, 2019). Notably, the African continent produced only around 4% of the world's total global greenhouse gas emissions in 2017. Nevertheless, due to topography, infrastructure and social conditions, African societies are among those 'most vulnerable to and least prepared for climate change. Even with external support, building climate resilience in Africa will take significant political will and resources, as well as support from populations that understand the need to prioritize climate action' (Selormey and Logan, 2019). Moreover, climate change literacy—the ability to keep informed not only about the reality of climate change but also about the reasons behind it—is not quite established, but in 2019 it did range at only 28% within a sample of 33 African countries (Selormey and Logan, 2019). A 2022 Afrobarometer survey on South Africa shows that 51.2% of the population still have not heard about the phenomenon of climate change, and 5.9% 'don't know' whether they are familiar with the concept (Hofmeyr, 2024). Compared to the previous Afrobarometer survey from 2021, climate illiteracy has increased significantly, as in 2021 only 43% had not heard about climate change at all (Mpako and Govindasamy, 2023). That means that in 2022, climate change awareness was only 42.7%, which is a decrease of 6–7% compared to the 2021 survey (Mpako and Govindasamy, 2023). This is an enormously high value, not only in statistical terms, that does not meet the expectations of global institutions and their agendas for global climate change governance. The latter was meant to significantly increase climate change awareness worldwide due to practices and education resulting from the SDG17 and to make the world more resilient.

These are significant numbers that emphasise the urgent need for climate communication in the Global South. However, what is also becoming clear is that climate change embeds itself into long-established realities of a global world order. When, in 2001, Michael Hardt and Antonio Negri introduced their approach of 'empire' as the form of global capitalist power within modernity, they were not aware of the meaning the challenge of globalisation would develop in only ten years; they even abandoned the concept of ecology completely. However, when they elaborate on the characteristics of modernity, they make an important point by saying that by implementing 'hegemony over the paradigm of modernity', what they call the 'forces of order' did not only 'reestablish ideologies of command and authority', but also deployed a 'new transcendent power by playing on the anxiety and fear of the masses, their desire to reduce the uncertainty of life and increase security' (Hardt and Negri, 2001: 75; Ahrens and Halbmayer, 2023: 2). The new global empire, according to Hardt and Negri's analysis, thus builds its strength on its capacity to connect to the modern virtues of creativity, agency, individualism, sovereignty and freedom, but at the same time it remains based on that new quality of a transcendent power. Today, it seems that climate change, as a rather uncannily intangible phenomenon in constant development, has taken the position of a contemporary transcendence. This could be the reason that it has developed so quickly into the new leading paradigm for global sovereignty and its governance. The only difference is that a planetary mode of sovereign governance can no longer respond to the yearning for more security and less uncertainty. On the contrary, climate change as a signifier of the modern transcendent establishes a logic of permanent crisis and insecurity. However, the present empire is driven by the concept of climate change as a globally effective signifier—much to the surprise of Hardt and Negri who have, in their neglect of ecology, radically underestimated the power of certain aspects of this concept.

One question that emerges from this situation, therefore, is to what extent the newly minted climate change governance bodies and agencies not only perpetuate an established order of global power but also shape its contemporary expression. If climate change defines a new global reality that merges social and ecological aspects and literally includes everybody everywhere, does it, then, also define the need for a redistribution of power on global scale? Another question is to what degree the challenge of global climate change translates into the social realities that establish themselves in the various global areas. Is the problem of global climate change consequently integrated on the local level of communities and lifeworld practices? If so, which were the specific local strands and capabilities considered that were sought to suit the integration of a phenomenon such as global climate change—which is, in fact, not an entirely new phenomenon anymore? Far from being mere minor aspects of a larger problem, these issues define how climate change is situated, perceived, thought and responded to within a greater variety of global topographies. It is therefore the discrepancies and ambiguities that are produced and made visible by the global nature of climate change that need to be addressed. First of all, it is significant that the most affected and most vulnerable countries are, without exception, located in the so-called Global South. In November 2023, the United Nations Economic Commission for Africa (UNECA) stated that of the 20 countries most threatened by climate change, 17 were in Africa (UNECA, 2023). Hence, Africa is significantly affected by climate change impacts, although the continent's share in the production of global emissions is still relatively low. Today, climate change impacts burden the GDP of many African countries and have implications for both 'economic and social development strategies' (UNECA, 2023).

For three basic reasons, it is, therefore, necessary to address the global phenomenon of climate change from a postcolonial perspective. Firstly, the emergence of climate change in particular parts of the global topography obviously triggers new developments of a world order of growing inequality. Even if this was a contingent result (the analysis of which would require more and deeper interdisciplinary research), it is remarkable how precisely the global mapping of climate change vulnerability and disadvantages matches with the cartography of the global world order, understood as Empire by Hardt and Negri, that follows the world order of (post)coloniality. In this light, their diagnosis that 'a new paradigm of power, a postmodern sovereignty' (Hardt and Negri, 2001: 138) might come to the fore. Quite in line with Luc Boltanski's and Eve Chiapello's analysis in *The New Spirit of Capitalism* (2005), they posit that a postmodern sovereignty replaces the well-established 'modern paradigm and rule through differential hierarchies of the hybrid and fragmentary subjectivities' (Hardt and Negri, 2001: 138). According to them, the irony here would be that these 'modern forms of sovereignty would no longer be at issue, and the postmodernist and postcolonialist strategies that appear to be liberatory would not challenge but, in fact, coincide with and even unwittingly reinforce the new strategies of rule' (Hardt and Negri, 2001: 138). This line of thought is worth following in a different context than its original, argument which does not at all refer to environmental issues. It seems that Hardt and Negri have not been aware of the essential qualities of this 'new paradigm of power'. With the emergence of climate change as a planetary problem, today, the category of nature has evolved into one of the core elements that make global power unfold in the present. Therefore, that new paradigm of power is specifically enforced via the successful implementation of transnational bodies and procedures of climate change governance that, by definition, operate on a global scale instead of a national one and that, at the same time, include strategies for development, modernisation and resilience. As much as Hardt and Negri have participated in the design of new materialism (Gamble et al., 2019) in the late 1990s with its renaissance of Marxian theory in a changed world due to globalisation and digitisation, this approach is still too focused on classic approaches to labour, value production, etc. In contrast, today, the global regime of power operates, to a great extent, via strategic attempts at an institutionalisation of response measures to the challenges imposed by nature that should have been mastered by the same rule of modern power that Hardt and Nagri still consider in charge, but transformed. That is, today, the subject of power transforms into that of nature and its expressions. Nevertheless, it also prolongs a well-established system of global postcolonial rule. Power and disempowering by nature thus seem to be the true new postmodern form of sovereignty, and it massively contributes to the further consolidation of global inequalities.

Secondly, and at first seemingly contrary to the first point, it increasingly turns out that the essential availability of basic natural resources makes a planetary common. As Shangrila Joshi has argued, the 'climate crisis warrants [. . .] an embracing of the practice of the commons' (Joshi, 2021: 6). Therefore, access to water, air, food, and all the other resources that form the essential basis of human existence needs to be secured and guaranteed. By focusing her research on the annual COPs as key institutions of a contemporary transnational climate change governance, Joshi argues that 'power dynamics occur at this site, and the ways in which power is distributed at this site impinges on power dynamics at particular local site' (Joshi, 2021: 12). This emphasises the argument that the challenge of global warming functions as a catalyst for the emergence of new forms of power and sovereignty. However, in line with such development there has been a change in the logic of the commons, as the new normal seems to have the quality of disaster as 'extreme weather events such as droughts, hurricanes, heatwaves, drier conditions enabling fires and floods are becoming more commonplace' (Satgar, 2018: 3). A just distribution and access to the basic commons necessary for human existence in the anthropocene, therefore, should be one of the key drivers of approaches to climate justice. To reflect on the questions of climate (in)justice means to acknowledge that 'global climate change is fundamentally about injustice and inequality—in vulnerability, responsibility, and mitigation, as well as participation in the global economy' (Roberts and Parks, 2007: 97). This, in turn, requires a new understanding of globalisation, because the social and institutional articulation of climate change has so far closely followed the logic of a (post) colonial divide of the modern world. Hence, climate justice assumes that the commons of human existence—water, air, food—first need to be acknowledged as such and that their establishment as levers for regulating global interaction requires both a new understanding of globalisation and of the notion of the planetary. As globalisation, in its classic and still valid sense, means the 'monopolization of the world's resources' (Urry, 2014: 53) according to the demands of production and consumption in the Global North, there can be 'no literal globalization' (Urry, 2014: 53). To this background, the understanding of climate change as a process that also fosters the meaning of the 'commons' (Ostrom, 1990; Engel and Saleska, 2005) as common goods for which access for everybody needs to be secured is twofold: As much as the commons are basically there and, therefore, need to be secured by social and political institutions, they, at the same time, only need to be invented and institutionalised by cultural practices of recognition, knowledge, and awareness.

Thirdly, all this integrates into the emergence of a new power regime for the contemporary that, as argued above, no longer relates to the classic categories of modernity and its critique, but rather leans on nature as the very resource that the logic of modernity and its urge for suppression has rendered vulnerable. Although the recent system of foundation and dissemination of power still leans on a more classic understanding of globalisation, it most prominently deploys its modes of domination via the regulation, governance and control of nature.

It is therefore not an accident that climate change has become the signifying phenomenon of the Anthropocene. It counts among the major worldwide challenges, and is virtually undisputed among academics and in transnational governance bodies. It impacts not only the environment, but from the point of view of human societies, it also severely affects the overall conditions of social life, its organisation and order. Hence, climate change has become a significant factor in the further development and governance of societies worldwide. It is for this reason that, as a globally emerging phenomenon that affects any place, person, or society on the planet, climate change has been correctly defined as a planetary phenomenon.

As outlined above, climate change seems to have taken the position of the new transcendent of modernity. On the one hand, this is due to its abstract nature which makes it impossible to experience it directly as it unfolds over longer timespans and in larger areas. These characteristics, however, allow it to serve as the modern transcendent by virtue of it being employed as a means of transnational governance and, therefore, a new mode of power. In their attempt to apply the category of the planetary in social thought, Nigel Clark and Bronislaw Szerszynski also refer to Hardt and Negri's approach, especially to their terminology of 'multitude'. To this end they introduce the term of 'earthly multitudes' to pinpoint 'the way that different human groups or collectives respond to the multiplicity that inheres in our planet' (Clark and Szerszynski, 2021: 9). Their suggestion is that this will make possible 'an understanding of this multitude as a shared way of responding to the challenges raised by the changeability of the Earth and the opportunities opened up by planetary self-ordering and variation' (Clark and Szerszynski, 2021: 9). Hence, they try to provide a counterstrategy that subverts the design of the planetary as a category for the justification of new forms of global power and control by the expansion of dispositifs of transnational governance, and to provide an understanding of the planetary as a gateway for counter practices. This is why they do not rely on an understanding of the planetary as a homogenous entity but rather see it as a 'multiplicity' (Clark and Szerszynski, 2021: 22).

In so doing, they intend to delegitimise the until now-established global dominance of the Global North. And they are making it quite clear that 'under conditions of a deep, rumbling and repressed anxiety over the abyssal ungrounding of the Earth, Western thought reimagines human difference in such a way that Europe's others will not only serve as the economic foundations of Earth-transcending accumulation; they will comprise a physical substrate that supervenes between 'full' humanity and a dangerous planet' (Clark and Szerszynski, 2021: 114). This is precisely where the postcolonial legacy of contemporary climate change action comes in. However, in spite of it being a part of a mostly obvious system of global injustice and inequalities that draws back on the baggage of colonial legacies, a postcolonial perspective needs to be applied in order to gain an understanding of the currently emerging system of transnational climate change governance as this system utilises postcolonial legacy in more subtle ways injecting it with new qualities. It is therefore necessary to analyse both dimensions of postcoloniality in global climate change agency, namely the rather obvious perpetuation of the long-established dominance by the Global North over the Global South and its transfer into a new and more abstract concept of planetary concern.

In its basic planetary dimension, climate change severely impacts food security and water accessibility, it influences processes of national and transnational migration, it alters the social as much as the material infrastructure and it shapes the increase and radicalisation of social inequality. Hence, as climate change further impacts on almost every dimension of social life, it turns into the most important driver of contemporary processes of sociation. It is for this reason that climate change and its socio-political consequences also raise, in new ways, questions about which social order and which political system should be not only reasonable but appropriate to meet the requirements of climate change mitigation as well as the demands of open and diverse society. Only by responding to such questions, societies—all of them, in one way or the other, based on ideas of modernity and modernisation—may be able to adequately address today's keenly felt threat of a change in the world precipitated by the impact of climate change.

Here, as is well known today, the Global South is especially affected, the Southern African region being one of those regions worldwide that is hit most severely. In general, the latest IPCC report says climate change vulnerability 'is higher in locations with poverty, governance challenges and limited access to basic services and resources, violent conflict and high levels of climate-sensitive livelihoods (e.g., smallholder farmers, pastoralists, fishing communities) (high confidence)' (IPCC, 2023: 101). Although, the report states, several risks could be moderated by adaptation measures, even this option is related to modes of global inequality.

> The largest adaptation gaps exist among lower income population groups (high confidence) and adaptation progress is unevenly distributed with observed adaptation gaps (high confidence). Present development challenges causing high vulnerability are influenced by historical and ongoing patterns of inequity such as colonialism, especially for many Indigenous Peoples and local communities (high confidence). Vulnerability is exacerbated by inequity and marginalisation linked to gender, ethnicity, low income or combinations thereof, especially for many Indigenous Peoples and local communities (high confidence). (IPCC, 2023: 101)

The concept of colonialism and its legacy in particular is repeatedly mentioned in the IPCC report as a marker of climate change vulnerability in the Global South (IPCC, 2023). Hence, coloniality is acknowledged as a factor for which societies are rendered particularly vulnerable. This perspective is emphasised by a variety of recent surveys. According to Afrobarometer, only 43% of the South African, 37% of the Mozambican, and 29% of the Botswana population have heard of climate change (Mpako and Govindasamy, 2023). In the Southern African region, climate change awareness is still rudimentary. Although the majority of citizens who are aware of climate change say it is making life in the country worse (Mpako and Govindasamy, 2023), active climate change policies in the region are underdeveloped. Concern for continuous development and procedures for modernisation is usually still in the foreground and also emphasised in many National Development Plans. There is a good reason for this, as vulnerability is tied to poverty and processes of marginalisation. Therefore, the inherent logic in many such Development Plans is that an increase in social safety and life quality also contributes to a reduction of environmental vulnerability. However, as approaches in post- development have pointed out, all too often, the understanding of development is not linked to improvements in the social condition but follows the strand of its economisation (Esteva, 2010). As has been often criticised, the understanding of modernisation itself is still based on an understanding of social development and the good life that consequently adopts the classic paradigm of modernisation as it has been historically invented by the European societies and implemented by the Global North. Its successful, lasting establishment in the Global South ignores the quality of domination, suppression, and global inequality that is inherently part of the development concept. Plus, it is the modern analogy of economic development and environmental extraction that has made the contemporary and future atrocities of climate change possible (Gaillard, 2021). The on global scale still prevailing assumption is that development necessarily proceeds in a straight line as a 'temporalizing concept' (Tomba, 2019: 21). In this concept, Tomba argues, 'different populations have different roles to play in the development of universal history and, finally, they are configured as different stages toward the final goal of a universal civilization' (Tomba, 2019: 21).

Thus, the irony is that the acceptance of the classic development paradigm guarantees that, due to its competitive nature and the different stages that different actors or societies take, the final goal of a universal civilisation characterised by the equal development of all its members will never be achieved. Instead, the temporalising concept of development will establish a hierarchy of development stages which, so far successfully, bars the so-called developing countries from accomplishing development. In this way, development itself is a placeholder for the continuation of postcolonial power. An alternative understanding of development would be required to meet the contemporary demands of global multitudes in the Anthropocene. It is from this perspective that not only the development of the Global South, but especially more recent concepts of sustainable development as first drafted by the 1987 Brundtland report to the UN, have been criticised from the angle of post-development for just renaming and prescribing a classic understanding of development (Esteva, 2010). Also, an emphasis on sustainability goals is far from the emergence of a new understanding of development, being still attached to economic growth and the commodification of nature and its resources (Sachs, 2010). Hence, the establishment of a postcolonial agenda of the contemporary that also responds to the various increasing environmental challenges is as complex as it is subtle, although in many ways it is also rather visible.

Against this broader background, the overall perception of the global reality of climate change as a gateway for a new kind of global equity in the Anthropocene seems questionable. It appears that the emergence of the planetary challenge of climate change produces a new quality of equal affectedness by disaster to which only the planet as a whole is able to respond via a cooperative effort. Remarkably, the fact that climate change occurs as a planetary factor that does not spare anybody is directly linked with the recognition of the human species as a geological force in the Anthropocene. Dipesh Chakrabarty (2012) draws the conclusion from this that in the light of climate change and the Anthropocene, we are forced to revalue the legacy of contemporary postcolonial critique. He argues that 'in becoming a geophysical force on the planet, we have also developed a form of collective existence that has no ontological dimension. Our thinking about ourselves now stretches our capacity for interpretive understanding. We need nonontological ways of thinking the human' (Chakrabarty, 2012: 13). When, according to Chakrabarty, such shared affectedness by impending disaster demands a non-ontology of the human species, his argument is that we need to abandon the anthropological anthropologies of cultural worldviews. Facing a planetary challenge like climate change, these anthropologies were manifested as obstacles to reasonable responses. The nonontological point of view, according to Chakrabarty (2012), does not mean a non-anthropocentric repositioning but the enabling of a shared experience of and a response to climate change as the human species. This is because 'we' as the human species need to respond to climate change through cooperation, which ultimately makes postcolonial concerns secondary as the planetary dimension of mutual affectedness is highlighted. This narrative of global equity in the face of climate change is, of course, itself burdened by problems, as neither climate change nor transnational institutions or measures have so far been able or willing to turn that planetary challenge into a new understanding of global equity that would appropriately meet the demands of the social transformations induced by climate change.

Although various institutions for climate change governance have been established in recent years—such as the 2015 Paris Agreement, the 1994 UNFCCC, the IPCC reports that have been issued since 1990, and probably most important, the annual COPs—they consistently follow a common logic that is mostly criticised as a rationality that in turn follows the epistemology, agenda and interests of the Global North. Even if this seems to be a simplification as it does not address the complexity of the interactions, differences and ambivalences in the global world, it seems clear that transnational climate change governance sustains the dominance and prevalence of the existing structures of global order. The open desperation with which UN representatives are currently fighting for climate change mitigation and by which they—since around 2023—blatantly signal the most likely failure of the UN Agenda of *Sustainable Development Goals* is not in line with the simple dichotomy of local grassroots actors and a postcolonial system of domination. More likely, contemporary global modernity operates according to Theodor W. Adorno's idea of the administered world (Adorno, 1990). An analysis of the logic of power in this administered world would show how modes of institutionalisation are constantly colonised by a rationality of bureaucracy that eventually only caters to itself and is detached even from the intention of its inventors,e.g., the UN. Clark and Szerszynski likewise point out that the reduction of 'the question of living on a volatile Earth to a matter of uneven social responsibility' (Clark and Szerszinsky, 2021: 101) is made too hastily and is itself rooted in concepts that no longer meet the demands of the present crisis.

It is within this logic of an administered world that the emergence of climate change contributes in many ways to the deepening of the divide between the Global North and the Global South. Existing dependencies, structures of domination and influence and positions of global hegemony are not coming to an end; they are not transformed into new practices and relations on a worldwide scale. Instead, the existing realities are often not only preserved but also radically deepened. Global and transnational politics and institutions demonstrate this problem again, most recently in November 2023 during the COP27 in Egypt, when the Global North actors resisted the provision of sufficient financial resources for the Global South and finally accepted it with great reluctance.

In essence, climate knowledge is brought about by a rationality that relies on the commodification of nature and extraction of natural resources and is dominated by a selection of privileged countries. The phenomenon of climate change itself is largely an effect of the production of carbon emissions by countries from the Global North. Their potential for industrial development has not least been made possible by exploiting colonised societies and extracting their resources. Thus, climate change is one of the later emerging results of 'the darker side of modernity' (Mignolo, 2011). However, categories like the Global North and Global South are not stable; they are permanently fluid, as China's rise as the most significant emitter in the world perfectly demonstrates. Nevertheless, China still manages to label itself as a developing country and an opponent to the Global North. The strategic nature of such categories, and their often poor analytical value, is thus becoming evident. South Africa's role in Southern Africa as by far the greatest emitter in the region and a leading Global South nation is just as ambivalent as China's role in Asia. Nevertheless, the United Nations Office for Disaster Risk Reduction (UNDRR, 2023) confirms that the ten countries affected most by climate change impact today are located in what we call the Global South, in the southern African region, and that they are almost exclusively low carbon emission producers.

Financial resources to make adaptation and mitigation measures possible are still largely held by the Global North, whose dissemination policies also define the globally effective options for further development. According to Walter Mignolo, the process of transformation of 'nature into natural resources in the West was a sign of progress and modernization and at the same time a sign that other civilizations stagnated and were falling behind the West' (Mignolo, 2011: 13). This procedure, he says, resulted in a strategy of coloniality to wrap up 'nature' and 'natural resources' into a 'complex system of Western cosmology' (Mignolo, 2011: 13). To this day, Mignolo concludes, the global order, or what Hardt and Negri call the 'empire', based on the epistemology that resulted from the process initiated by the Industrial Revolution, is being exported around the globe by the system of colonialism. Mignolo's conclusion is that decolonisation and dewesternisation are the same process more or less, the 'end of a long history of Western hegemony' (Mignolo, 2011: 48), especially regarding their respective 'epistemic standards' (Mignolo, 2011: 49).

What Mignolo calls for is a sort of emancipation from Western domination not only in material structures and social and political systems but also in terms of consciousness and practices of the mind. The quest for a new epistemology, or more broadly, a new ontology, is the quest for the foundation of a new epistemological identity and the break with one that is considered to be burdened with colonial atrocities. The question is whether this will be possible. Is a complete detachment of the postcolonial from its former colonial matrix and from the legacy of Western thought (which is not necessarily the same thing) possible at all? Arguing for nothing less than postcolonial sovereignty, Olúfémi Táíwò warns of the acceptance of 'false binaries like "the West" (an idea) versus "Africa" (a place)', which would lead the postcolonial societies 'directly to nativism and atavism' (Táíwò, 2022: 130). In the case of global climate change, the question is what this all means. In how far does global climate change serve as a catalyst for the prolongation and strengthening of well-established structures of maginalisation and suppression of postcolonial topographies worldwide? What would be an appropriate strategy for response to climate change that goes beyond the established Western reasoning? How is the postcolonial articulated in the vulnerable societies, and what are its consequences?

Against this background, this book considers it essential to reflect on the postcolonial signature in climate change approaches, knowledge, communication and measures, with particular emphasis on the Southern African region. This book is divided into several parts, each focusing on a different aspect of climate change from a transdisciplinary postcolonial perspective, drawing on social sciences and the humanities. First, it focuses on the change in the governmental approach to climate change in a peculiar postcolonial context from a historical perspective (Tempelhoff). Second, the chapters question the pitfalls of the dominant ontology of modernity and ask what alternatives should be considered (Coetser, and Pressend). Third, the epistemological approaches to climate change knowledge are discussed and analysed by focusing on postcolonial contexts in Southern Africa (Terblanché-Greeff, Alm, and Faimau). In conclusion, attention is drawn to the establishment of a postcolonial climate change regime and what it may entail (Ahrens).

Johann Tempelhoff (in Chap. 2 of this volume) provides a historical analysis of a particular postcolonial context by offering insight into the evolving nature of South Africa's relationship with water management and climate change issues. This analysis is set against the backdrop of 'how the transformation of statecraft paved the way for current climate change awareness in postcolonial South Africa' (Chap. 2). In the 20th century, there was a shift from a focus on water scarcity to a focus on climate change influencing water scarcity. However, this shift was gradual.

In the 1950s, South Africa became a member of the World Meteorological Organisation (WMO), and scientists of the South African Weather Bureau (SAWB) 'maintained sound relations in research and international science and data sharing' (Chap. 2). During this period, climate change increasingly featured on the scientific research agenda in the USA. Despite South African scientists' participation in the global scientific discourse, the government did not shift its primary focus from water issues to climate change awareness per the global trend. 'In 1961 the [National Party] government [of South Africa] set its own agenda for "decolonisation" and the transition to a republic, ruled by the white minority' (Chap. 2). This agenda was exemplified by the government's policy of separate development, which later became known as 'apartheid'. As the decolonisation of other African countries gained momentum in the 1960s, South Africa became increasingly authoritarian, which resulted in sanctions in various domains. The SAWB became isolated from the international scientific community. By 1975, their membership of the WMO was suspended, and all formal ties with the UN were broken. This meant that SAWB 'was severed from international data systems and knowledge on new emerging climate change science' (Chap. 2). The government's primary focus on water governance remained, and this 'became evident in the government's foregrounding of the environmental dimensions of water in the state's communication services' (Chap. 2).

The release of Nelson Rolihlahla Mandela from jail in 1990 paved the way for political deliberations. As a result, South Africa 'rejoined the mainstream international activities of the UN, and its delegates were once again welcomed and participated at the UN Rio Earth Summit of 1992' (Chap. 2). In 1994, the SAWB resumed its full membership of the WMO. 'In the country's water sector, there were early signs of awareness of environmental matters and climate change' (Chap. 2). However, because of its political isolation, 'new ideas and advanced technologies were initially slow to find their way into all institutional governance structures' (Chap. 2). Nonetheless, South Africa's meteorologists are currently 'active participants in deliberations on global climate change, planning and research' (Chap. 2). Contemporary South Africa is actively committed to international collaborations on climate change matters as per the 2015 Paris Agreement.

Yolandi Coetser (in Chap. 3 of this volume) states that the climate crisis can be attributed to what she refers to as a problematic 'view-of-nature' on which the contemporary human-nature relationship is based. This specific view-of-nature—historically salient in the Western paradigm—sees nature as predominantly having instrumental value, which is a key contributor to anthropogenic climate change in the Anthropocene. Fundamental to this view-of-nature is the superiority complex of humans. She provides a historical narrative of this relationality by focusing on key philosophical figures, trends and epochs that have initiated and perpetuated a dualistic worldview that separates humans from nature, ultimately justifying its domination. She then explores the interconnectedness of patriarchy, colonialism and anthropocentrism as issues that originate from a shared 'logic of domination' (as posed by Warren, 1990). This logic is rooted in Western intellectual history and grounded in oppositional value dualisms. Here, women, subalterns and nature are seen as the *Other*. That is, they are marginalised, depoliticised and rendered invisible.

Ecofeminism is proposed as a promising framework to address the problematic view-of-nature as it highlights the interconnected oppression of women, subalterns and nature. By employing the concept of postcolonial justice, she argues there is a need to examine the common denominators of coloniality and anthropogenic climate change. This, however, will not suffice because 'postcolonial justice must also consider the patriarchal sentiment that underlies much of the Western view of nature' (Chap. 3). In a nuanced manner, ecofeminists argue that the root of the problematic view-of-nature is not primarily anthropocentric but androcentric. This worldview is associated with traditionally masculine values like domination, exploitation and control that have shaped patriarchal human-nature relationality based on the man/nature dualism. Coetser argues that, all things considered, it is appropriate to implement a postcolonial ecofeminist approach when striving for environmental justice in the Anthropocene. Specifically, such an approach recognises the shared oppression of the *Other* and seeks to dismantle the underlying dualistic thinking in favour of a more nuanced worldview.

Michelle Pressend (in Chap. 4 of this volume) unpacks aspects of Western modernity's ontology as 'rationalist, reductionist and rooted in Cartesian dualism' (Chap. 4). She turns her attention to the mainstream Anthropocene discourse, which, she argues, is still overreliant on techno-scientific and market-orientated approaches grounded in Western modernity's dualistic ontology of separation. These approaches (to climate change response), such as sustainable development and 'green growth', still operate within 'a singular story of development, progress and growth' (Chap. 4) where nature is seen as a resource. That is, the hegemonic logic of exploitative systems remains at play, albeit under different guises in global neoliberalism. To illustrate, she posits that postcolonial nation-states of the Global South demand redress for climate change as it is predominantly caused by the anthropocentric behaviour of developed countries. However, such demands (e.g., just energy transition and energy democracy) are based on Western modernity's ontology as they advocate for an integration into the established paradigm of progress and development. This paradigm encompasses social, political, and economic praxis associated with modernisation and human mastery over nature. As a result, redress is created in the image of the Western 'view-of-nature', which is based on capitalist logic.

However, it becomes clear that the 'master's tools will never dismantle the master's house' (Chap. 3). As such, alternative approaches stemming from outside the neoliberal constructs of the Anthropocene discourse should be investigated. It is essential to recognise and acknowledge non-dualist relational ontologies 'that are in defence of social production of life, and how these values and ways of being might assist in responding differently to the climate crisis' (Chap. 4). It is then a pluriversal approach to addressing climate change that is appealed to. That is an approach that invites multiple ways of being-in-the-world, being-with-others, and knowing that are not in line with the dominant Western paradigm (representative of the 'one world story'). The *Other*—marginalised groups such as women and indigenous people—that have once been silenced through domination can now facilitate the necessary shift needed to cultivate regenerative and life-affirming solutions to the climate crisis that stretches beyond the 'heteropatriarchal, white supremacist and colonialist social orders prevalent in the world today' (Chap. 4) to reimagine human-nature relationality. Pressend's key intention is to bring to the fore the need for alternative approaches that can decenter the dominant dualist ontology (and the resultant problematic view-of-nature) while opening up space for non-dualist ontologies, particularly those found in indigenous cosmologies and non-Western worldviews of those perceived as the *Other*.

Aïda Terblanché-Greeff (in Chap. 5 of this volume) contends that international climate change governance institutions, like the IPCC, create and disseminate climate change knowledge from a Western scientific paradigm. National governments of developing countries often use such knowledge received from the Global North in their climate change adaptation strategies and policies. Ideally, these strategies and policies should be nuanced and context-specific. However, there is an overreliance on the techno-scientific approach to climate change knowledge rooted in the Western modernisation ontology (marked by scientism), which leads to the marginalisation and exclusion of indigenous and local knowledge systems in climate change discourses in some developing countries. She argues that this is because not everyone is created equal as epistemic subjects regarding climate change knowledge. That is, '[s]ome knowers are considered to be superior to others' (Chap. 5). This specific form of the logic of domination leads to epistemic oppression and, ultimately, epistemic injustice. Like other forms of injustice, there is a power imbalance in epistemic injustice caused by historical harm patterns, such as patriarchy and colonialism (grounded on value opposition dualisms found in Western ontology as illustrated by Coetser in Chap. 3 and Pressend in Chap. 4).

Epistemic injustice is perpetuated in some postcolonial societies through long-standing discriminatory socio-economic structures and persistent linguistic barriers. For example, in contemporary South Africa, the continued dominant use of English in government climate change policy documents excludes many rural communities whose members might not be proficient in the *lingua franca* but who possess valuable indigenous and local knowledge that can contribute to the development of context-specific approaches to addressing climate change. As a result, some groups (viz. Africans) that were once considered the *Other* (when colonised) are *still* wronged in their capacity as epistemic subjects of climate change knowledge. Addressing climate change knowledge epistemic injustice in a postcolonial context will, however, require more than the mere inclusion of marginalised voices in climate discourses. Instead, the process of decolonisation of climate change knowledge should be facilitated by challenging systemic structures of inequality and fostering inclusive dialogue through the acknowledgment of diverse epistemologies. Furthermore, the decolonisation of climate knowledge 'would involve substantial shifts in the way climate change mitigation and adaptation strategies and policies are created and communicated' (Chap. 5) in a postcolonial context like South Africa.

Elisabeth Alm (in Chap. 6 of this volume) states that the political system of Botswana 'consists of two juxtaposed parts, the democratically elected, modern system implemented after Botswana gained independence from the United Kingdom in 1966 and the traditional indigenous leadership system [known as *kgotla*], which was already in place during pre-colonial times' (Chap. 6). At face value, the decentralised government of Botswana provides a model for including diverse epistemologies in public climate change discourse and governance. This system of government acknowledges the importance of incorporating indigenous knowledge *and* modern scientific knowledge into climate strategies and policies. Alm analyses the interaction between the modern democratic structures of the Botswana government and the *kgotlas* on climate change governance, policies and strategies. *Kgotlas*—headed by a chief—operate on a village level and aim to create a forum for information dissemination and community input on various topics, e.g., climate change. Furthermore, they provide consultation on national and local governance. The multi-level governance interaction is motivated by the fact that the *Kgotla* has legitimisation and strong communal ties, thus offering the Botswana government opportunities to provide localised climate governance. However, 'the incorporation of indigenous leadership [of *kgotlas*] in governance is confronted with certain hardships' (Chap. 6) in this particular postcolonial context. For example, 'local climate change governance in Botswana faces the duality of having a decentralised political system with the *kgotla*, a space often described as well-equipped to take part in local governance, while simultaneously operating in a postcolonial world order that still perpetuates wilful hermeneutic ignorance and therefore undervalues the expertise offered by the *kgotla*' (Chap. 6).

Although Botswana's climate policies and strategies emphasise the importance of incorporating indigenous and local knowledge into climate governance, Alm has critiqued the tangible implementation as tokenistic. To overcome this tokenism, real integration is required. That is, moving beyond mere performative inclusion (as illustrated by Alm in Chap. 6) towards actually empowering communities (at village level) by valuing their epistemic contributions to climate change governance.

Gabriel Faimau (in Chap. 7 of this volume) starts by presenting decoloniality as a school of thought. As part of the decolonial turn, discourse primarily focuses on 'interrogating the hegemony of the Western knowledge systems, unmasking the Western epistemological tradition and reframing the quest for knowledge production in the context of a pluriversality of epistemic traditions and cultures'

(Chap. 7). Furthermore, decolonial scholarship 'has focused on the discourse of decoloniality to examine the dark side of epistemology that has dominated the epistemic experiences of the Global South' (Chap. 7). The problems of Western modernity's power structures and the resultant hegemonic epistemology have also been identified and diagnosed. In response, decoloniality promotes epistemic liberation as it acknowledges multiple epistemologies of the *Other* that exist regardless of the dominance of the Western scientific episteme. However, a 'clear proposal on how to bring indigenous voices, knowledges or epistemologies to the centre of scholarship and research" is still lacking (Chap. 7).

Faimau formally introduces the concept of 'decolonial reflexivity' as a practical operationalisation of decoloniality. It is a 'strategic positioning in understanding decoloniality, particularly in the context of the relationships between and among cultures' (Chap. 7). To start, decolonial reflexivity acknowledges that colonial epistemology still 'dominates the interpretation of knowledge systems and produces a structure of authority that governs knowledge production' (Chap. 7). In response to this acknowledged hegemony, decolonial reflexivity takes a dialogue-orientated approach to knowledge co-production. It perceives'epistemological design and knowledge production as a conversational arena for mutual exchange among various epistemologies that reflect the particularities of different societies and cultures' (Chap. 7). Through this notion of knowledge co-production, decolonial reflexivity can facilitate epistemic justice. This is exemplified when knowledge co-production undertakes a collaborative process whereby various groups bring together a plurality of epistemologies to create new knowledge, which can be used to address a communal problem, such as climate change. By broadly examining climate change research in Botswana, Faimau explores how decolonial reflexivity has been and can be further applied to the co-production of climate change narratives.

The IPCC has a legacy of being the leading institution of epistemological climate change governance. Its reports are seen as the 'gold-standard' of climate change science, which stems from the fact that one of the main purposes of the IPCC is to 'provide global, consistent, and unequivocal knowledge about climate change' (Chap. 8). This is done by reviewing existing climate change knowledge and subjecting it to academic and authoritative judgment. Jörn Ahrens (in Chap. 8 of this volume) focuses on the IPCC's knowledge production practices from a postcolonial angle. He questions what it means when knowledge about climate change—as a global phenomenon—is based on the hermeneutical resource pool of those with epistemic power under the guise of universalism. This resource pool (of the Global North) is grounded in the techno-scientific episteme of modernity.

In contemporary society, it is argued that climate change is a 'pressing global challenge that needs to be confronted transnationally through co-ordinated action of multiple actors worldwide' (Chap. 8). By implementing a planetary universalist approach, the IPCC promotes the unilateralisation of climate change knowledge to facilitate a shared understanding and support a global climate change governance regime to address this challenge. However, this approach 'can only be one that neglects difference by the implementation of a meta-perspective that melts down any aspect of *otherness*' (Chap. 8). The standardisation of 'knowledge as one coherent epistemological resource thus significantly contributes to the production and emergence of global inequalities as epistemic injustice' (Chap. 8). This results in epistemological monopolisation and postcolonial epistemic hegemony that suppresses non-Western epistemologies of the Global South by ignoring 'existin-gexpertise, experiences, and knowledge [from] many regions worldwide' (Chap. 8). IPCC's one-world approach counters the pluralisation of knowledge as needed in various postcolonial contexts.

Another effect of planetary universalism is the notion that climate change is a universal risk that affects all of humanity equally. This neglects the regional differences in degree of vulnerability and responsibility, ultimately removing the political dimensions which have to be considered in climate justice. This is in line with another purpose of the IPCC, as it aims to stay a politically unbiased and scientifically objective participant in the global climate change discourse. By doing this, the IPCC 'manages to remain the unquestioned technocratic expert on climate change, the most important political consultant within the transnational discourse and among the bodies of climate change governance' (Chap. 8). Thus, it is argued that the IPCC's work contributes to the overall formation and maintenance of a postcolonial climate change regime despite its intent to provide objective scientific assessments.

References

Adorno, T. W. (1990). Kultur und Verwaltung [culture and administration]. In T. W. Adorno (Ed.), *Gesammelte Schriften*: Vol. 8. Soziologische Schriften I. (3rd ed.). Frankfurt am Main: Suhrkamp

Ahrens, J., & Halbmayer, E. (2023). Epistemologies of global warming in the South. The social and cultural dimension of climate change in Southern Africa. In J. Ahrens, & Halbmayer, E. (Eds.), *Climate change epistemologies in Southern Africa: Social and cultural dimensions* (p. 1–23). London & New York: Routledge. https://doi.org/10.4324/9781003180814

Boltanski, L., & Chiapello, E. (2005). *The new spirit of capitalism*. London & New York: Verso

Chakrabarty, D. (2012). Postcolonial studies and the challenge of climate change. In *New Literary History*, 43(1), 1–18

Clark, N., & Szerszynski, B. (2021). *Planetary social thought: The anthropocene challenge to the social sciences*. Cambridge, UK, & Medford, MA: Polity Press. https://doi.org/10.1515/ngs-2021-0036

Engel, K. H., & Saleska, S. R. (2005). Subglobal regulation of the global commons: The case of climate change. In *Ecology Law Quarterly*, vol. 32, 2/2005, 183–233

Esteva, G. (2010). Development. In W. Sachs (Ed.) *The development dictionary: A guide to knowledge and power*. London & New York: Zed Books

Gaillard, J. C. (2021). *The invention of disaster: Power and knowledge in discourses on hazard and vulnerability*. London: Routledge

Gamble, C. N., Hanan, J. S., & Nail, T. (2019). What is new materialism? In *Angelaki. Journal of the Theoretical Humanities*, Vol. 24, 6/2019, 111–134

Hardt, M., & Negri, A. (2001). *Empire, Cambridge, MA, and London*. UK: Harvard University Press

Hofmeyr, J. (2024). Summary of results: *Afrobarometer Round 9 survey in South Africa, 2022*. Compiled by the Institute for Justice and Reconciliation. Retrieved September 11, 2024, from https://www.afrobarometer.org/wp-content/uploads/2024/03/Summary-of-results-South-Africa-Afrobarometer-Round-9-7mar24.pdf

IPCC (2023). Climate Change 2023: Synthesis report. Contribution of working groups I, II and III to the sixth assessment report of the Intergovernmental Panel on Climate Change [core writing team, H. Lee, & J. Romero (Eds.)]. IPCC, Geneva, Switzerland. https://doi.org/10.59327/IPCC/AR6-9789291691647

Joshi, S. (2021). *Climate change justice and global resource commons: Local and global postcolonial political ecologies*. Routledge.

Marrumbe, D. (2024). Cancellation of UN climate weeks removes platform for worst-hit communities. *Climate Home News*. 28/03/2024. Retrieved July 8, 2024, from www.climatechangenews.com

Mignolo, W. (2011). *The darker side of western modernity: Global futures, decolonial options*. Durham & London: Duke University Press

Mpako, A., & Govindasamy, P. (2023, March 9). *Despite growing evidence, climate change is still unknown to many South Africans*, no. 165. Retrieved July 11, 2024, from https://www.afrobarometer.org/wp-content/uploads/2023/03/AD615-Climate-change-still-unknown-to-many-South-Africans-Afrobarometer-8mar22.pdf

Ostrom, E. (1990). *Governing the commons: The evolution of institutions for collective action*. Cambridge: Cambridge University Press

Roberts, J. T., & Parks, B.C. (2007). *A climate of injustice: Global inequality, North-South Politics, and climate policy*. Cambridge, MA, & London: The MIT Press

Sachs, W. J. (2010). Environment. In Sachs, W. J. (Ed.), *The Development Dictionary: A Guide to Knowledge and Power*. London & New York: Zed Books

Satgar, V. (2018). The climate crisis and systemic alternatives. In Satgar, V. (Ed.), *The climate crisis: South African and global democratic eco-socialist alternatives*. Johannesburg: Wits University Press

Selormey, E. E., & Logan, C. (2019, September 23). African nations are among those most vulnerable to climate change. *A new survey suggests they are also the least prepared.* Retrieved August 7, 2024, from https://www.afrobarometer.org/articles/african-nations-are-among-those-most-vulnerable-climate-change-new-survey-suggests-they-are/

Táíwò, O. (2022). *Against decolonisation: Taking African agency seriously*. London: Hurst and Company.

Tomba, M. (2019). *Insurgent universality: An alternative legacy of modernity*. New York, NY: Oxford University Press.https://doi.org/10.1093/oso/9780190883089.001.0001

United Nations (2023). *RCCs: Accelerating implementation.* United Nations framework convention on climate change. Bonn: UN Campus

UNDRR (2023, March 20). *10 countries at risk of climate disaster*. Retrieved November 9, 2024, from https://www.preventionweb.net/news/10-countries-risk-climate-disaster

UNECA (2023, November 2). *17 out of the 20 countries most threatened by climate change are in Africa, but there are still solutions to this crisis*. Retrieved July 12, 2024, from https://www.uneca.org/stories/17-out-of>-the-20-countries-most-threatened-by-climate change-are-in-africa%2C-but-there-are

Urry, J. (2014). *Climate change and society*. Cambridge & Malden, MA: Polity Press

Warren, K. J. (1990). The power and promise of ecological feminism. *Environmental Ethics*, 12(2), 125–146

Postcolonial South Africa's Water and Aridity in the Era of Climate Change

2

Johann Tempelhoff

Abstract

South Africa's long walk from the Union of South Africa, a colonial-type state established in 1910, to a liberal democracy more than 80 years later, is evident in the way it perceived water scarcity and dealt with the emergent crises of water scarcity and regional aridity. For the greater part of the century, the state's climate and water experts were reluctant to acknowledge climate change. Because of the country's racial policies, South Africa was increasingly marginalised in the realm of international politics from the 1950s onwards. Two decades later, South African weather scientists were excluded from participating in the activities of the World Meteorological Organisation (WMO) where climate change was in focus. In the 1970s, developing countries of the Global South increasingly shaped international politics and collaboration. The South African state was then isolated because of its apartheid policies. In a number of critical fields South African science and technology was in danger of falling behind in ever-modernising international trends. Meteorology was one such field. Only after the country made the transition to a liberal democracy in the 1990s were the country's scientists able to rejoin WMO activities and United Nations (UN) deliberations on increasingly important global climate change conditions.

J. Tempelhoff (✉)
North-West University, Potchefstroom, South Africa
e-mail: Johann.Tempelhoff@nwu.ac.za

A. C. Terblanché-Greeff et al. (eds.), *Climate Change and the Postcolonial*, Kulturelle Figurationen: Artefakte, Praktiken, Fiktionen,
https://doi.org/10.1007/978-3-658-49328-8_2

Keywords

Climate change · Postcolonial development · Water · Global North · Global South

Introduction

Historically, large parts of Southern Africa[1] have been shaped by aridity and an ever-present vulnerability to water scarcity. The subcontinent is home to the Namib—one of the world's oldest deserts, dating back 50–66 million years (Ollier, 1977; Logan, 2021). The interior regions of Southern Africa have an average annual rainfall of 500 millimetres per annum (mm/a). They extend over a vast area from Cape Town's Table Mountain, up South Africa's West Coast and interior, into Namibia, the southern Angolan coastal region, eastwards through Botswana and parts of Zimbabwe, as far as 230 km off Mozambique's Indian Ocean coastline (Lovegrove, 2021). For many centuries, these sandy areas have shaped Southern Africa's climate conditions and its valuable water resources (Watermeyer, 1905).

In this chapter, the focus is on water, aridity and a growing awareness of climate change in the 20th-century South Africa. Although fiercely debated, climate change has certainly shaped governance and meteorology of drought conditions in this water-stressed subcontinent Until mid-century British, German and Portuguese colonial rule prevailed and the region's indigenous peoples were subject to political, social and economic ecological governance by colonial European minorities.

After World War II, in the spirit of the United Nations' (UN) recognition of universal human rights, decolonisation picked up the pace in Global South regions where indigenous peoples aspired to achieve liberation. Against this backdrop the focus falls on how the transformation of statecraft paved the way for current climate change awareness in postcolonial South Africa.

[1] The region includes the 16 member states of the Southern African Development Community (SADC): Angola, Botswana, Comoros, Democratic Republic of Congo, Eswatini, Lesotho, Madagascar, Malawi, Mauritius, Mozambique, Namibia, Seychelles, South Africa, United Republic of Tanzania, Zambia, and Zimbabwe.

Current Climate Change Projections for Southern Africa

In 2021, the sixth report of the United Nations' Intergovernmental Panel on Climate Change (IPCC) focused on global regional trends in climate change and water-related matters. Southern Africa is now more vulnerable to climate-related disasters (IPCC, 2021). In contrast to the Northern Hemisphere, where warming temperatures are bound to cause more rainfall, Southern Africa will be subject to increased temperatures and diminishing and erratic rainfall events (Engelbrecht, 2021).

The IPCC (2021) explains that the western parts of Southern Africa are warming up and are destined to become drier and more arid. In the eastern parts of the subcontinent rainfall events are predicted to become more erratic and scattered.[2] This holds profound implications for the subcontinent's finite water resources. A 2023 report of the World Meteorological Organisation (WMO) suggests that in the period 2002–2022 there were measures in place to secure terrestrial water storage facilities in Southern Africa. But in South Africa there were signs of lapses in water supply, and these tended to become more prevalent (WMO, 2023).

The fact that Southern Africa, and especially South Africa, feature prominently in the IPCC report, is important. Clearly, the country's meteorologists are active participants in deliberations on global climate change, planning and research. There is also evidence of an active South African commitment to international collaboration in matters of climate. This is encouraging and speaks of mutual trust between local climate specialists and their colleagues elsewhere in the world. But this has not always been the case.

In the 1970s, South Africa was forced to withdraw as a member of the WMO—the leading international meteorological organisation responsible for globally co-ordinating meteorological activities. Countries of the Global South, supporting the liberation struggle on the subcontinent, were at the forefront in calling for South Africa's withdrawal. This meant that the South African Weather Bureau (SAWB) was severed from international data systems and knowledge on new emerging climate change science. Only with the onset of South Africa's multiracial transition to democracy in the 1990s did the newly named South African Weather Service (SAWS) rejoin the WMO.

The 2021 IPCC report is thus of substantial symbolic relevance. It reflects the fact that the country is not only an influential Global South state but is also again part of international initiatives on vitally important current and future meteorological trends.

---

[2] For more information, refer to the regional fact sheet (IPCC, 2021).

There was a time when South Africa was excluded from participation in international deliberations on meteorological trends. But since the 1990s the country has regained access to international talks on climate change. South Africa's renewed participation in the current global climate change science inspires confidence.

Union of South Africa: Water Governance and Climate

In 1910 the Union of South Africa became one of the first independent states in Southern Africa. The subcontinent was notable for its European characteristics in which Portugal, Britain and Germany held sway over indigenous peoples. From the outset the Union was ruled by a white minority government, pursuing many former colonial policies favouring predominantly European-styled 'white' urban communities. Its indigenous peoples were confined to under-developed residential spaces on the urban outskirts, with limited water and sanitation infrastructure systems.

The Union of South Africa was on a rapid growth path. The country was a prime exporter of minerals such as diamonds, as well as gold and farming produce. Industry, commerce and the country's collective economy grew at a pace. Towns and cities attracted people from elsewhere in Africa and overseas seeking employment or opportunities for entrepreneurship. Rapid growth entailed rapid increases in water consumption in the household sector, industry and agriculture. As water had been a scarce resource in the region for a very long time to begin with, droughts and intermittent floods had devastating consequences, and water security became vital for the smooth running of the state and the well-being of its population.

Within two years of its establishment the Union Parliament approved the *Irrigation and Conservation of Waters Act* of 1912 (U of SA, 1952). In the spirit of the new law, the government prioritised irrigation farming strategies steeped in the British Imperial tradition (Strange 1906). Because of the country's aridity new irrigation technologies were introduced. An important benchmark was irrigation technology developments in the arid western United States of America (Lewis, 1915).

The 'conservation' component of the country's water legislation was also intended to 'restore' South Africa's natural lush mountain forest lands and their river catchment areas. However, by the 19th century many water sources had started drying up because of the over-exploitation of indigenous forests (King, 1938, 1941), in the eastern and southern Indian Ocean coastal regions (Bennett and Barton, 2018).

In many parts of the country reports of drought conditions suggested that attention had to be paid to the country's finite and diminishing water resources. Political leaders and officials became more aware of aridity and water scarcity after the former German colony of South West Africa (Namibia) became a mandated territory of the Union of South Africa after World War I. The enlarged arid region provided for a transboundary consolidation of scientific and governance strategies aimed at mitigating drought and flood conditions in the arid regions of Southern Africa.

In 1912, the Union's Department of Irrigation (DoI) established a meteorological division. In addition to constant assessments of available water resources, officials recorded data on climate, floods and drought conditions. In 1925 the government established the (SAWB) that actively engaged with all government departments on matters of meteorological relevance (King, 1959).

The country's growing thirst for water was often brought up in feisty parliamentary debates in the 1910s. Droughts posed a persistent 'wicked problem' for the government (Vogel et al., 2016). Providing well-managed water supplies to arid regions was of critical importance, and it required meteorological expertise and more useful data. After World War I (1914–1918), the former German colony of South West Africa (Namibia) also became a mandated territory of South Africa (Wessels, 2017). Insight gained on a larger surface area of aridity in Southern Africa enabled the Union's hydrological and meteorological experts to make better water conservation management plans and strategies for dealing with arid conditions larger parts of the subcontinent.

Drought Investigations and Plans

In 1917, the government appointed a drought investigation commission to examine much-publicised reports of climate and 'water problems'. Droughts caused by destructive anthropogenic over-consumption of natural resources became harbingers of change in arid parts of the country. The commission explained that droughts were caused by overgrazing and land degradation. Overused soil had dried out and grazing lands seldom managed to recover (Department of Agriculture [DoA], 1926).

The country's farmers and rural communities remained sceptical of the government experts, and local home-spun views proliferated on matters of weather, climate and water. Many of these views were shaped by the 19th-century colonial missionaries who reported on apparent 'changing climate conditions' in the drier parts of the subcontinent (Moffat, 1843; Livingstone, 1858).

Subsequently, many farmers even called on the authorities to consider 'greening the desert'—a suggestion made in the 1870s by the Cape Colonial Botanist (Brown, 1875).

Based on scientific knowledge and comprehensive insights on aridity in different territories, the commission described the public views as 'inappropriate' (DoA, 1926). For the commission it was about water and not the land. Farmers were advised to focus on their available 'units of water', instead of the 'units of land' they used (DoA, 1926). The drought commission in neighbouring South West Africa (Namibia) expressed similar views, and the farmers there were also advised to opt for livestock farming. Mixed farming practices were not considered viable in desert areas (Botha, 2005).

In the public realm there was no shortage of new ideas on how to deal with water scarcity. In 1918, E. H. L. Schwarz started a public debate on the 'desiccation of Africa' (Plug, 2020). As a former Union geologist and lecturer in geography at Rhodes University, Schwarz argued that the arid Karoo region had been part of a water-rich subcontinent as recently as 'three centuries' earlier. Since then, it had simply 'dried out' (Schwarz, 1920). He recommended a strategy for 'watering the desert', i.e., by combatting the 'crisis' of aridity in the western and north-western parts of Southern Africa (Alcock, 2010; Plug, 2014). His thinking was informed, in part, by 19th-century German science, articulated by Ferdinand Gessert as well as by missionaries in German South West Africa (Siiskonen, 2015; Plug, 2017). It was symptomatic of the views shaped by the plans for British and German colonial expansion in Southern Africa (McKittrick, 2015).

Schwarz proposed that—besides existing natural lakes—a set of artificial inland lakes be created to 'distribute rainfall' to dry regions in the subcontinent's interior. The 'lakes', of which Namibia's ancient natural Etosha Pan was one, were intended to become artificial reservoirs filled by rivers flowing from the Angolan highlands, Bechuanaland (Botswana) and the Namibian interior towards the Atlantic. Schwarz explained that the Atlantic coastal winds carrying daily flows of mist along the coastal desert areas inland would increase and drive rain clouds into the interior of Namibia and South Africa (Schwarz, 1920).

In 1925, a team of leading experts on water and arid conditions embarked on an expedition to explore at first hand the areas demarcated by Schwarz (DoI, 1926). His plan subsequently came under criticism. In addition to the national water department, the government's drought commission explained that there were too many uncertain factors and insufficient data to make accurate calculations on Schwarz's theory. F. E. Kanthack, a former head of the Union's Irrigation Department, attended a public meeting in London, where Schwarz presented a paper on his scheme. Kanthack castigated Schwarz's theory in a scathing article (Special Correspondent, 1926) to quell the enthusiasm for British colonial expansion.

Climate Change Views Before the 1950s

Despite much talk about drought and flood events, meteorologists and hydrologists still had no scientific certainty about anthropogenic climate change in Southern Africa prior to the 1950s. Since the 19th century there had been several key scientific discoveries in the field (IPCC, 2007), with more following in the 20th century. But scepticism was prevalent; in South Africa, climate and water experts remained wary of climate change talks.

In 1934, T. E. W. Schumann, chief meteorologist of the Union, hinted at 'changing climatic conditions', measured in terms of localised rainfall data going back to the late 1800s. He did not favour cyclical rainfall theories, but conceded that there were secular variability patterns because of changes in local rainfall conditions (Schumann, 1934). Another government expert, Reenen van Reenen, a US-educated engineer and first chairperson of the Union's Irrigation Commission, was aware of cooling oceans and changing climate conditions related to El Niño-Southern Oscillation (ENSO) events, first described in the 1870s. Van Reenen used ENSO observations to identify comparative trends along the coastlines of South America's Pacific and South Africa's Atlantic ocean (Van Reenen, 1935). However, he made no overt statements on climate change.

Even in the national water department (DOI), despite the fact that the country's water sector had a vested interest in climate issues, there was hardly any research on climate change. In 1942, D. F. Kokot, a senior departmental engineer and researcher, even argued against climate change. He had studied the concept based on the views of contemporary meteorologists and hydrologists. Although he correctly identified the role of carbon dioxide and greenhouse gases, in line with earlier hydrological investigations, he argued that changes in climate were cyclical and could at best be predicted in terms of astronomical sun spot theories developed in China and dating back to approximately 400 BCE (Bod, 2022). Sun spots were considered useful as pointers because they were said to have occurred at fairly regular intervals, It could lead to tracking similar cycles (Kokot, 1942). In his PhD thesis submitted at the University of Pretoria, Kokot explained that in his view, 'climatic change' could not be used to account for drought and flood conditions in the management of the country's water resources (Kokot, 1948).

Post-World War II South Africa and Climate Change

As a founding member of the UN, South Africa found itself in a well-connected international community. The SAWB was exemplary in terms of international science and technology networking. Since the end of World War II, knowledge systems on international science and technology spread rapidly across the globe. In January 1950 South Africa became a member of the newly formed WMO. At the time of the 1953 WMO regional conference in Tananarive, South Africa's delegates were appointed to lead a sponsored research project to compile a climatological atlas of Africa (Anon., 1954). Throughout the 1950s, the SAWB and WMO maintained sound relations in research and international science and data sharing (King, 1959).

In the 1950s, climate change featured increasingly on the agenda of scientific research in the USA. There were concerns about solar radiation, the warming of the oceans, the impact of industrial activities and the effect of pollution on water resources (Ashford et al., 1990). But nuclear defence strategies in the bipolar world of the capitalist West and communist East inhibited international collaboration on global science strategies. (Hart and Victor, 1993). Consequently, the WMO initiated a global atmospheric research programme under the auspices of the UN. The objective was to develop global insights on climate conditions (Demeritt, 2001).

While American physicists focused on 'rapid climate change' processes (Weart, 2003), civil society activism was on the rise in the 1960s USA. This was part of the environmental awareness of a new post-World War II generation of 'baby boomers' (Hart and Victor, 1993; O'Riordan, 1999). Environmental activism became increasingly popular on university campuses in the West. The scientist and activist, Rachel Carson, became an iconic role model through her studies on how water pollution in the US endangered aquatic ecosystems and human communities (Carson, 1962). However, in South Africa, environmental awareness and civil society activism was far more subdued.

South Africa Responds to Early Decolonisation

After World War II there were significant political shifts in South Africa. In 1948 the National Party (NP) came to power. Its policy of separate development, later labelled Apartheid, focused on a discriminatory, racialised style of governance, one which foregrounded the ethnic identities of the country's indigenous African communities. In a rapidly developing modern authoritarian state, the NP government wanted black South Africans to move to their traditional Homelands, with the prospect of 'their own' future 'sovereign independence'. There were even plans to confine first nation descendants of indigenous Late Stone Age San and Khoi peoples in separate areas. Apart from being at odds with the UN's Universal Declaration on Human Rights, the South African state, across a broad spectrum of activities, veered away from long-standing international relations and conventions of engagement at the onset of Africa's era of decolonisation.

In 1955, the NP government faced severe criticism when members of several anti-apartheid organisations met in Kliptown, south of Johannesburg, and deliberated on a freedom charter, calling for the scrapping of Apartheid and the normalisation of a non-racial South African society. The event reverberated internationally within the UN network, where several member states monitored and became highly critical of racial issues in the country (UN, 1955). The South African Government became more security conscious and ramped up its spending on defence. In 1957, Ghana broke free from British colonial rule and became the first African state to proclaim a republic. Several others followed soon.

Controversial South African Government policies and Britain's exclusion of South Africa from the British Commonwealth subsequently culminated in the Sharpeville Massacre in March 1960, an event that reverberated in many parts of the world (Clark and Worger, 2023). In 1961, the NP government set its own agenda for 'decolonisation' and the transition to a republic, ruled by the white minority (Devenish, 2013).

Determined not to bow to international condemnation, sanctions and escalating domestic upheaval, the NP government survived for several decades, partly because of the demand for its mineral resources (Manson and Mbenga, 2003). However, as Africa's decolonisation gained momentum the country became increasingly isolated and more authoritarian. The struggle led by the African National Congress, the Communist Party and the Pan Africanist Congress was gaining ground.

In addition to the focus on food security, the government's water strategy foregrounded the ever-growing demand for energy. Generating electricity and manufacturing petroleum products from coal required vast amounts of water. The government focused on a more effective system of water distribution in many parts of the country, based on ground-breaking water research and extensive construction projects.

Advances in South Africa's Water Management

The advances in water research and the harnessing of the two major rivers in central South Africa, the Orange and the Vaal, were impressive and deserve the highest praise. Water experts designed complex transfer systems to provide water to urban areas, manufacturing industries, mining as well as the country's irrigation farming areas (Pegram and McKenzie, 1991; Murray, 2023). However, despite these impressive developments, it is also true that up to the 1990s, residents of African townships and the country's rural areas seldom had sufficient access to water and sanitation as rapid urbanisation increased the infrastructure development backlogs.

The paramount importance of water in South Africa became evident in the government's foregrounding of the environmental dimensions of water in the state's communication services. International environmental issues and growing civil society activism remained secondary to the apartheid state's preoccupation with water.Consequently, primarily European-inspired nature conservation initiatives for the country's fauna and flora proliferated. Most prominent were nature conservation NGOs like the World Wildlife Fund (WWF) and the Southern African Nature Foundation (SANF). However, the threat of civil society activism was nipped in the bud. One ray of hope at the time was the onset of environmental research and teaching programmes at several South African universities (Steyn and Wessels, 1999).

The only evidence of environmental awareness in the 1970s was a comprehensive report by a group of experts working on a strategy for the development, conservation and growth of the country's water resources into the new millennium (RP34/1970, 1970). The move was a ploy, according to water experts in academia. Des Midgley from the University of the Witwatersrand was critical of the government's call for the celebration of a national Water Year, while the rest of the world celebrated environmental awareness under the auspices of the UN (Midgley, 1971).

Global South and Global North

For the greater part of the Cold War Era (1948–1990), developing countries of the future Global South were referred to as 'Third World' countries on account of their allegiances to either the West, in the North Atlantic Treaty Organisation (NATO), or the Communist East's Warsaw Treaty countries. Several developing countries were able to pursue non-aligned policies for their own benefit and frequently stymied the plans made by larger states.

In terms of identity, the Global South's self-conception gained momentum in the 1950s as newly formed states grouped together at conferences to share views on decolonisation and statehood. The 1955 Bandung Conference in Indonesia was the first of its kind. China, India and Indonesia were at the forefront of promoting collaboration amongst the leaders of the new developing states in Africa, Asia, Latin America and the Middle East. They met in Indonesia to deliberate on their role in the international community. Soon several developing countries, like Yugoslavia, Egypt, India, Ghana and Indonesia were amongst the first to be grouped as 'non-aligned states' of the world. it was a ploy of the emergent post-colonial small countries to refrain taking sides in the pro-capitalist (West), or pro-Communist East. In the process they wielded more political muscle, by favouring the globe's new post-colonial countries.

In the 1960s there were occasional references to the developing countries in the southern regions of the world (Heine, 2023) as the concept of Global North and Global South co-operation gained traction. Leading diplomats and politicians actively promoted plans for capitalist Western countries to engage with the leadership of the developing countries. At the forefront of this initiative were the United States, as well as UN-affiliated institutions such as the World Bank (WB), the World Trade Organisation (WTO) and the Organisation for Economic Co-operation and Development (OECD). In the capitalist West there were strong views on the need to engage with developing countries.

In the mid-1970s, Willie Brandt, a former West German Chancellor, accepted an invitation to lead an independent commission on international issues which included leaders of developing countries. The Brandt Report was submitted to the UN in 1980 (Anon., 1980), and it recommended sweeping changes to the international economy and a better understanding of difficulties experienced by former colonial territories, as independent states, in a globalising environment where Western capitalism could play a significant role (Mesbahi, 2006).

At the time, the world was still caught up in a Cold War environment, but then China, under the new leadership of Deng Xiaoping, broke ranks with other communist states in 1978 to follow a semi-capitalist growth path in reviving a Chinese spirit of political moderation and responsible statehood, after the dictatorship era of Mao Zedong (Pye, 1993). For Global North states, it strengthened plans for reshaping engagements with independent developing countries of Africa and the rest of the Global South. At the time, developing countries of the Global South played an increasingly influential role by supporting the frontline states of Southern Africa.

South Africa faced imminent isolation. Primarily for security reasons, NATO still supported the SAWB's membership in the WMO, because of the country's strategic location at the southern tip of Africa (Jones et al., 1969). However, at the sixth congress of the WMO in Geneva in April 1971, the representatives of several countries were unhappy about South Africa's participation. SAWB officials reported back that most countries still had 'no qualms' with South African presence (Bosua, 1971).

The global energy crisis after the 1973 Yom Kippur War in the Middle East and an embargo on fuel exports from the Middle East wreaked havoc in the highly industrialised economies of the West. A global economic recession set in. In Southern Africa this coincided with heightened conflict and greater demands for the liberation of former colonial territories. Although the High Commission territories of Lesotho (1966), Botswana (1966) and Swaziland/Eswatini (1968) had gained independence, many Southern African colonies had not yet done so.

South Africa's government, quite aware that the country had reached peak gold production (Feinstein, 2005), stepped up its efforts on upgrading its production of synthetic liquid fuel from coal to protect itself. Consequently, the country's comprehensive river water transfer systems, especially that of the Vaal River, were upgraded to generate more energy produced from coal for electricity and petrochemicals for petroleum in the Eastern Highveld regions of South Africa (Anastasi, 1980), which had significant implications for increasing greenhouse gas emissions causing climate change.

In 1974, following a military coup in Portugal, Angola (1974) and Mozambique (1975) gained their independence (Sidaway and Simon, 1993). These new Southern African frontline states increasingly displayed a distinct postcolonial style in advocating for the liberation of all the subcontinent's former colonial territories (Willems, 2013).

On the international science and technology front, support for decolonisation also made headway. Countries of the Global North made substantive investments in science and technology partnerships with former colonial states. Their voices were also heard in deliberations. In 1974, the UN's Special Committee on Apartheid reported that South Africans attending a WMO conference on the International Hydrological Decade had been asked to leave. Their presence was said to be harmful to the anticipated success of the conference (UN, 1974). The announcement came shortly after the coup in Portugal and the independence of Mozambique and Angola.

Besides the frontline states, the Organisation of African Unity (OAU) also targeted South Africa and objected to its mandate of administrating South West Africa (now Namibia). The Cold War between the East and the West intensified on the subcontinent, and the primary goal was to apply more pressure on apartheid South Africa (Azania) and Rhodesia (now Zimbabwe) where white minority governments were embroiled in warfare against freedom fighters. The Rhodesian Government declared a 'bush war' while the South African Defence Force started operations in a 'border war' on Namibia's border with Angola and neighbouring states.

Similar trends reverberated in the UN's international scientific sector. On 30 April 1975, at the fourth plenary session of the seventh WMO Congress, South Africa's membership of the WMO was suspended. A total of 66 representatives of member states voted in favour of the suspension. A further 25 countries voted against the measure while 14 abstained (WMO, 1975; South African History Online, 2019). At the time, formal ties between the UN and South Africa were broken (Steyn, 2018).

The Cold War escalated in Southern Africa. Cuba's military support for Angola and the OAU call for the liberation of Namibia forced the USA to get involved in talks with the white-minority governments of Zimbabwe and South Africa. The US's Secretary of State Henry Kissinger was eager to promote 'majority rule' in Zimbabwe and 'self-determination' in Namibia (Vale, 2023). White rule in South Africa after the 1976 Soweto uprisings and the ongoing ANC struggle against the apartheid state increasingly pointed to the need for negotiations to end Apartheid and the need for Namibia to gain independence. A major breakthrough was when Rhodesia (Zimbabwe) opted for black majority rule in 1980.

South Africa and Neighbouring States in the 1980s

Despite the intensified international struggle against apartheid South Africa, some remnants of social and economic interdependence remained in some Southern African states. For example, in 1986, an agreement was reached between apartheid South Africa and the Kingdom of Lesotho for the development of the transboundary Lesotho Highlands Water Project (LHWP) to secure greater supplies of water for distribution on the subcontinent (Davies, 1994). This agreement formed part of a more comprehensive interdependent regional water security strategy.

For the economy of the Kingdom of Lesotho, the LHWP meant the inflow of much-needed revenue from water sales and access to hydro-power electricity in a mountainous rural country. The plan was not without its opponents. The agreement was concluded at a time when international civil society activism stepped up protests against large dam schemes that negatively affected socio-ecological conditions in many parts of the world (Goldsmith and Hildyard, 1986). Despite South Africa's isolation, and thanks to the country's continued international collaboration on matters of dam construction, essential precautions were in place. Local and international engineering specialists and companies working on the LHWP soon embarked on significant environmentally friendly infrastructure development strategies, long before the LHWP's completion in 2003 (Van Vuuren, 2013).

The LHWP was born in an era of growing global environmental consciousness which began with the UN's 1972 Stockholm Conference on the Human Environment (UN, 1972) and culminated in 1987 with the release of the UN's Brundtland Report on the Environment (titled *Our Common Future*) which calls for sustainable development (United Nations Environment Programme [UNEP] 1987). Strong international lobbying groups foregrounded issues of climate change (Ashford et al., 1990). In Europe and the United States environmental organisations started exchanging ideas. Leading the field were the Climate Action Network (CAN), the WWF, Environmental Defense Fund, as well as Greenpeace International (GI), which helped in setting up the influential Climate Action Network (CAN) (Nulman, 2015).

Environment, Climate Change and South Africa's Transition

By the late 1980s, 'green politics' were used to develop mutually beneficial strategies in international relations between the Global South and North. American diplomats in particular were engaged with their European counterparts on topics related to the environment. In the transition to a post-Cold War world, issues of this nature took centre stage in Europe as the political thaw of Cold War thaw set in.

In July 1989, US President George Bush Snr visited Europe for a G7 summit with leaders of the world's wealthiest nations. He also talked to leaders of Eastern European states, many of whom were in the process of breaking ties with the Union of Soviet Socialist Republics (USSR). Some even attended a 'Green Summit' coinciding with the bicentennial celebrations of the French Revolution in France. About global warming, delegates were told that the newly established IPCC would see to climate change matters being dealt with to the satisfaction of all member states (Marz et al., 1989).

In South Africa, the release of Nelson Mandela in February 1990, after 26 years of incarceration, paved the way for political deliberations leading up to the country's first multiracial democratic elections in 1994. South Africa thus rejoined the mainstream international activities of the UN, and its delegates were once again welcomed and participated at the UN Rio Earth Summit of 1992. They also participated in WMO plans to enlarge the membership of the IPCC—the first concerted global effort to reduce carbon and greenhouse gases emissions (Roddick, 1997).

In the country's water sector, there were early signs of awareness of environmental matters and climate change. In 1990, South Africa's Water Research Commission (WRC) indicated its interest in funding climate change research (WRC, 1990). Roland Schulze of the University of KwaZulu Natal, who had been working on climate, hydrology and agriculture since the 1970s, published an outline of climate change and the hydrological implications of the anticipated greenhouse effect resulting from increasing carbon (CO_2) emissions into the atmosphere (Schulze, 1990). His long-time colleague Peter Tyson of the University of the Witwatersrand was involved in extensive modelling of long-term climate change conditions in Southern Africa (Tyson, 1993).

Post-Apartheid South Africa swiftly transitioned from an authoritarian to a culturally vibrant postcolonial liberal democracy in the 1990s. Of importance was the voice of an articulate and enthusiastic set of civil society organisations engaging with environmental and governance issues across racial lines. The number of NGOs grew rapidly, and there were also significant shifts towards public participation in the formulation of the government's environmental and conservation policies (Cock and Fig, 2001).

In the country's water sector, the national Department of Water Affairs and Forestry (DWAF) was slow to make the transition to climate change hydrology (Ashton et al., 1999). DWAF's hydrology and water resources management and development essentially relied on empirical data dating back to the mid-20th century. At the time, computer technology shifted from mainframe systems of the 1970s to personal computers in the 1980s. By the 1990s, thanks to increased computing power, skilled engineers, hydrologists, mathematicians and physicists expediated the shift towards water resource systems analysis and management. New computer databases provided capacity to develop stochastic modelling systems, simulating hydrological conditions more severe than those on record. These developments were conducive to future climate change research (Tempelhoff, 2018).

On 9 June 1994, shortly after South Africa's first non-racial democratic elections, the WMO at its 46th session of the executive council agreed that the SAWB could resume its full membership of the WMO. The staff bulletin announced that the SAWB was 'overjoyed by this historic decision and looking forward to participate fully in the activities of WMO' after almost two decades (Anon., 1994). This was followed by a symbolic name change to the South African Weather Service (SAWS).

Because of South Africa's political isolation, new ideas and advanced technologies were initially slow to find their way into all institutional governance structures. The DWAF formally mentioned climate change and the country's water resource management framework in its 2009 annual report (RSA DWAF RP163, 2009). DWAF stalwarts were steadfast in maintaining existing strategies for dealing with floods and droughts. But breakthroughs on climate change research in academia soon took place.

By the 2000s, more comprehensive and globally integrated databases and modelling systems emerged with hydrologists and meteorologists collaborating on forming more integrated modelling systems of knowledge on meteorology, hydrology and climate change (Engelbrecht et al., 2009; Engelbrecht, 2011). At the time, South Africa's countrywide water infrastructure systems showed signs of deterioration caused by overexploitaition of water resources, inadequate wastewater treatment services and overconsumption of strategically storaged water supplies.

In addition to evident climate change, reasons for growing water-related crises included ageing infrastructure systems, urban population growth, countrywide municipal service delivery protests and the first collapse of the electricity supply grid in 2007 and 2008 (Rondganger, 2008), compounded by the 2008 global financial crisis (Swatuk and Cash, 2018). Nonetheless, the government's policy of a principled, just transition to greater equality in terms of access to water resources services played a key role in water supply decision making (Jacobs et al., 2014).

Postcolonialism and Climate Change

When the Cold War ended in 1990 there was no longer any need for the 'Third World' developing countries of the Global South to remain marginal role players in the global economy. However, in some quarters a pervasive postcolonial mindset persisted. Memories of the postcolonial crises in developing countries were foregrounded in historical recollections and the search for new identities.

There were significant shifts in academia at the time. The international fixation on 'Western civilisation' in many intellectual discourses diminished as postcolonial studies accentuated different approaches to modernisation theory, global politics and economics. Unique cultures and modes of civilisation from around the world were acknowledged as distinctly different from those in Western

Europe. Experts in global studies aspired to reach consensus on compatibility and co-existence of cultures and societies in a globalising world (Conrad, 2016).

Climate change became a powerful driver of creating cohesion and collaboration amongst the countries of the world. In 2015, 196 countries attending a Conference of the Parties (COP) in Paris committed to maintaining the global average temperature at 1.5°C above the pre-industrial levels. However, in the face of the worsening climate change trends, scientists warned that temperatures were rising too fast, which could have serious ramifications.

They were correct. By 2019, a third of Africa's riverine areas were stressed and polluted with pathogens (Cousin et al., 2019). Southern Africa's finite riverine water resources and impoundment infrastructures were already at risk. The arid parts of the region were identified as vulnerable to water stress and diminishing food production, as was the case in the rest of Africa (Ayugi et al., 2022).

In 2023, UN Secretary General Antonio Guterres called on all countries to bring their net zero plans up by one decade. It was important to avert the danger of global heating above 1.5°C the average global temperature, of mid-19th century global industrial temperature (Anon., 2023). By 2023, the Earth's average surface temperature rose well above the previous 2016 record of 1.48°C (Voosen, 2024). Scientists even warned of critical tipping points if the planet's temperature were to rise by as much as 3°C in the 2030s (Armstrong McKay et al., 2022).

Conclusion

South Africa's long road to an integrated climate change science approach since the 1990s is the story of a modernising postcolonial African state where there is an ever-present awareness of the implications of water stress. It is also aware of anticipated future trends. Besides aridity, there is evidence of increasing cyclonal disaster events, especially in the southwestern Indian Ocean region. Intermittent periodic cyclones have wreaked havoc in the coastal and interior areas of the subcontinent. South Africa's Presidential Climate Change Panel (PCC, 2023), on the advice of the WWF, called for comprehensive disaster management plans and international collaboration to mitigate these threats collectively.

There is evidence in Southern Africa of an African connectedness in seeking international avenues of collaboration on science and effective strategies aimed at addressing issues of critical importance for the current and future development and sustainable growth of the subcontinent and its peoples. In the spirit of an African postcolonial mindset, the voice of South

Africa in international talks on climate change makes an important statement on active international collaboration on an issue that is no longer confined to anthropocentric views of the future. It has become a matter of planetary importance.

References

Alcock, P. (2010). *Rainbows in the mist: Indigenous knowledge, beliefs and folklore in South Africa*. Pretoria: South African Weather Service

Anastasi, J. L. (1980). *Sasol: South Africa's oil from coal story – Background for environmental assessment*. Final report Jul-Aug 79. US Environmental Protection Agency. (p. 44) EPA-600/8-80-002

Anon. (1954). Climatological atlas of Africa. *South African Journal of Science*, 51(4), 97–98. https://hdl.handle.net/10520/AJA00382353_303

Anon. (1980, March) Briefing paper: The Brandt Commission. London: Overseas Development Institute (ODI). Retrieved August 7, 2024, https://cdn.odi.org/media/documents/6638.pdf

Anon. (1994). *SA Weather Bureau back in WMO as full member*. Pretoria: SAWeerburo Nuusbrief

Anon. (2023). UN climate report: Scientists release 'survival guide' to avert climate disaster, News. *BBC World News*. Retrieved April 3, 2024, https://tinyurl.com/2ks6etr6

Armstrong McKay, D., Staal, A., Abrams, J., Winkelmann, R., Sakschewski, B., & Loriani, S. (2022). Exceeding 1.5 °C global warming could trigger multiple climate tipping points. *Science*, 377(6611). https://doi.org/10.1126/science.abn7950

Ashford, O., Gupta, S., Meade, P., Taha, H., & Weiss, G. (1990). *Climate change: Forty years of progress and achievement: A historical review of WMO*. Geneva: Secretariat of the World Meteorological Organization

Ashton, P., Braune, E., Maaren, H., McKenzie, R., Orpen, W., & Pitman, W. (1999). Hydrological science in South Africa: 1995–1998. *South African Journal of Science*, 95(6), 259–268

Ayugi, B., Eresanya, E., Onyango, A., Ogou, F., Okoro, E, & Okoye, C. (2022). Review of meteorological drought in Africa: Historical trends, impacts, mitigation measures, and prospects. *Pure and Applied Geophysics*, 179(4), 1365–1386. https://doi.org/10.1007/s00024-022-02988-z

Bennett, B. M., & Barton, G. A. (2018). The enduring link between forest cover and rainfall: A historical perspective on science and policy discussions. *Forest Ecosystems*, 5. https://doi.org/10.1186/s40663-017-0124-9

Bod, R. (2022). *World of patterns: A global history of knowledge*. Baltimore: Johns Hopkins University. https://doi.org/10.1086/725930

Bosua, T. (1971). Verslag van die bywoning van die 6e kongres van die WMO te Genève van 5 tot 30n April 1971. *Nuusbrief van die Suid-Afrikaanse Weerburo*. (pp. 80–83). Pretoria: SA Weather

Botha, C. (2005). People and the environment in colonial Namibia. *South African Historical Journal*, 52(1), 170–190. https://doi.org/10.1080/02582470509464869

Brown, J. (1875). *Hydrology of South Africa; or details of the former hydrographic condition of the Cape of Good Hope, and of causes of its present aridity, with suggestions of appropriate remedies for the aridity*. London: Henry S King

Carson, R. (1962). *Silent spring*. Boston: Houghton Mifflin

Clark, N., & Worger, W. (2023). *Voices of Sharpeville: The long history of racial injustice*. London & New York: Routledge. https://doi.org/10.4324/9781003257806

Cock J., & Fig, D. (2001). The impact of globalisation on environmental politics in South Africa, 1990–2002. *African Sociological Review/Revue Africaine de Sociologie*, 5(2), 15–35. https://doi.org/10.4314/asr.v5i2.23189

Conrad, J. (2016). *What is global history?* Princeton & Oxford: Princeton University Press

Cousin, E., Kawamura, A., & Rosegrant, M. (2019). The threat of water scarcity. From Scarcity to Security. *Chicago Council on Global Affairs*, 12–25. Retrieved January 22, 2023, from https://www.jstor.org/stable/resrep21409.6

Davies, A. G. (1994). An introductory description of the Lesotho Highlands Water Project, emphasizing implementation of phase IA. *Tunnelling and Underground Space Technology*, 9(1), 73–78. https://doi.org/10.1016/0886-7798(94)90012-4

Demeritt, D. (2001). The construction of global warming and the politics of science. *Annals of the Association of American Geographers*, 91(2), 307–337. https://doi.org/10.1111/0004-5608.00245

Department of Agriculture (1926). The great drought problem of South Africa. Pretoria: Government Printer

Department of Irrigation (1926). *Report of the Kalahari reconnaissance of 1925 with 5 plates and 29 figures*. Pretoria: Government Printer

Devenish, G. (2013). Cutting the apron strings: The South African experience of decolonisation. *The Journal for Transdisciplinary Research in Southern Africa*, 9(2), 309–340. https://doi.org/10.4102/td.v9i2.209

Engelbrecht, F. (2011). Projected regional climate futures for southern Africa: A changing planet – the atmosphere. *CSIR Science Scope*, 6(1), 8–10

Engelbrecht, F. (2021, August 17). Public talk: IPCC report: Implications for Africa and South Africa. *Meeting of climate justice charter movement, EFS, Wits and COPAC*. University of the Witwatersrand: COPAC. Retrieved August 2, 2024, from https://www.ipcc.ch/report/ar6/wg1/downloads/factsheets/IPCC_AR6_WGI_Regional_Fact_Sheet_Africa.pdf

Engelbrecht, F., McGregor, J., & Engelbrecht, C. (2009). Dynamics of the conformal-cubic atmospheric model projected climate-change signal over southern Africa. *International Journal of Climatology*, 29(7), 1013–1033. https://doi.org/10.1002/joc.1742

Feinstein, C. (2005). *An economic history of South Africa: Conquest, discrimination and development*. Cambridge: Cambridge University Press

Goldsmith, E., & Hildyard, N. E. (1986). *The social and environmental effects of large dams; case studies*. Camelford, UK: Wadebridge Ecological Centre

Hart, D., & Victor, D. (1993). Scientific elites and the making of US policy for climate change research, 1957–74. *Social Studies of Science*, 23(4), 643–680. https://doi.org/10.1177/030631293023004002

Heine, J. (2023, July 3). *The Global South is on the rise – but what exactly is the Global South? The Conversation.* Retrieved April 3, 2024, from https://tinyurl.com/2d85e2ma

IPCC (2007). Climate Change 2007 – The physical science basis. In M. Manning (Ed.), *Fourth assessment report of the Intergovernmental Panel on Climate Change. Intergovernmental Panel on Climate Change* (p. 996). Cambridge

IPCC (2021). Climate Change 2021: The Physical Science Basis. *Contribution of Working Group I to the Sixth Assessment Report of the Intergovernmental Panel on Climate Change: Regional Fact Sheet – Africa.* In Press V. Masson-Delmotte, A. Zhai, S. Pirani, L. Connors, C. Péan, S. Berger, et al. (Eds.). Retrieved April 3, 2024, from https://bit.ly/3xQOfzK

Jacobs, I., du Plessis, M., Trollip, K., & van Vuuren, K. (2014). *South Africa's 20-year journey in water and sanitation research.* Pretoria: Water Research Commission

Jones, W., Kruls, H., & Gretton, P. (1969). *NATO and South Africa.* South African Institute of International Affairs: SAIIA

King, J. (1959). The South African Weather Bureau. *South African Journal of Science*, 55(9), 227–232. Retrieved April 3, 2024, from https://hdl.handle.net/10520/AJA00382353_622

King, N. (1938). Historical sketch of the development of forestry in South Africa. *Journal of the South African Forestry Association*, 1(1), 4–16

King, N. (1941). The exploitation of the indigenous forests of South Africa. *Journal of the South African Forestry Association*, 6(1), 26–48. https://doi.org/10.1080/03759873.1941.9631098

Kokot, D. (1942). Irrigation possibilities in the Union of South Africa. *Civil Engineering*, 40, 166–219

Kokot, D. (1948). An investigation into the evidence bearing on recent climate changes over Southern Africa. *Irrigation Department Memoir.* Pretoria: Department of Irrigation

Lewis, A. D. (1915). Irrigation and settlement in America. *I. Department.* Pretoria: Government Printing and Stationary Office

Livingstone, D. (1858). *Missionary travels and researches in South Africa: Including a notebook of sixteen years' residence in the interior of Africa, and a journey from the Cape of Good Hope to Loanda on the West Coast: Thence across the continent, down the river Zambesi, to the eastern ocean.* New York: Harper & Brothers

Logan, R. (2021, September 2024). *Namib. Online Encyclopedia Britannica.* Encyclopedia Britannica, Inc. Retrieved June 2, 2023, from https://www.britannica.com/place/Namib

Lovegrove, B. (2021). *The living deserts of Southern Africa.* Cape Town: Struik Nature

Manson, A., & Mbenga, B. (2003). The Richest Tribe in Africa: Platinum-Mining and the Bafokeng in South Africa's North West Province, 1965–1999. *Journal of Southern African Studies*, 29(1), 25–47. https://doi.org/10.1080/0305707032000060598

Marz, l., Thomas, R., Warner, M., Meyer, M., & McDaniel, A. (1989). Bush's high-wire act: From Eastern Europe the 'Green' summit. *Newsweek*, 30, 8–11

McKittrick, M. (2015). An empire of rivers: The scheme to flood the Kalahari, 1919–1945. *Journal of Southern African Studies*, 41(3), 485–504. https://doi.org/10.1080/03057070.2015.1025339

Mesbahi, M. S. (2006). The Brandt Report. Share the World's Resources. Retrieved April 15, 2024, from https://sharing.org/information-centre/articles/brandt-report

Midgley, D. (1971). The impact of the water'70 campaign. *South African Journal of Science*, 67(4), 269–275

Moffat, R. (1843). *Labours and scenes in Southern Africa*. New York: Robert Carter

Murray, T. E. (2023). *Water for a thirsty land: The story of the Orange River Project*. Creda Press: Foot Print Press

Nulman, E. (2015). *Climate change and social movements: Civil society and the development of national climate change policy*. London: Palgrave Macmillan UK

O'Riordan, T. (1999). From environmentalism to sustainability. *Scottish Geographical Journal*, 115(2), 151–165. https://doi.org/10.1080/14702549908553824

Ollier, C. (1977). Outline geological and geomorphic history of the central Namib Desert. *Madoqua*, 1977(3), 207–212

PCC (2023, May). A critical appraisal of South Africa's just energy transition investment plan. *Recommendations from the Presidential Climate Commission p 29*. Retrieved April 3, 2024, from https://tinyurl.com/yryr3hd9

Pegram, G., & McKenzie, R. (1991). Synthetic streamflow generation in the Vaal River System Study. *Civil Engineering*, 33(1), 15–24

Plug, C. (2017). Gessert, Mr Ferdinand. *S2A3*. Retrieved January 22, 2017, from http://bit.ly/2jOSNEU

Plug, C. (2020). Schwarz, Mr Ernest Hubert Lewis (geology, geography, palaeontology). *S2A3 Biographical Database of Southern African Science*. Pretoria: S2A3. https://bit.ly/3RZAaw5

Plug, C. (2014). Schwarz, Mr Ernest Hubert Lewis *(geology, geography)*. Retrieved August 4, 2016, from http://bit.ly/2b5qXNZ

Pye, L. (1993). An introductory profile: Deng Xiaoping and China's political culture. *The China Quarterly*, 135, 412–443

Roddick, J. (1997). Earth Summit north and south: Building a safe house in the winds of change. *Global Environmental Change*, 7(2), 147–165 https://doi.org/10.1016/S0959-3780(96)00007-6

Rondganger, L. (2008, March 28*). Three months of load shedding*. Star

RP34/1970 (1970). *Report of the commission of enquiry into water matters*. Pretoria: Government Printer

RSA DWAF RP163 (2009). *Annual report: 1 April 2008–31 March 2009* (p. 390). Pretoria: Department of Water Affairs and Forestry

South African History Online (2019). *SA is suspended from World Meteorological Organization*. Retrieved June 19, 2023, from https://tinyurl.com/2atbqvbk

Schulze, R. (1990). Climate change and hydrological response in southern Africa: heading towards the future. *South African Journal of Science*, 86(7), 373–381. https://hdl.handle.net/10520/AJA00382353_4502

Schumann, T. (1934). Secular variations in South African rainfall. *Civil Engineering = Siviele Ingenieurswese*, 1934(1), 98–116. https://hdl.handle.net/10520/AJA10212019_14909

Schwarz, E. (1920). The Kalahari and its possibilities. *Journal of the Royal African Society*, 20(77), 1–12

Schwarz, E. (1920). *The Kalahari or thirstland redemption*. Cape Town: T. Maskew Miller

Sidaway, J., & Simon, D. (1993). Geopolitical transition and state formation: The changing political geographies of Angola, Mozambique and Namibia. *Journal of Southern African Studies*, 19(1), 6–28. https://doi.org/10.1080/03057079308708344

Siiskonen, H. (2015). The concept of climate improvement: Colonialism and environment in German South West Africa. *Environment and History*, 21(2), 281–302

Special Correspondent. (1926). The waters of the desert. *The Sun & Agricultural Journal of South Africa*, 17, 66–79

Steyn, P. (2018). Apartheid South Africa's participation in United Nations-Organized International environmental initiatives in the 1970s: A Reassessment. *Journal of Contemporary History*, 53(4), 861–884.https://doi.org/10.1177/0022009416678918

Steyn, P., & Wessels, A. (1999). The roots of contemporary governmental and non-governmental environmental awareness in South Africa, 1654–1972. *New Contree*, 45, 61–81

Strange, W. L. (1906). Irrigation and inter-colonial co-operation. *Transvaal Agricultural Journal*, 5(17), 1–16

Swatuk, L., & Cash, C. (2018). Perspectives on the Nexus: Water, Energy and Food Security in an Era of Climate Change. In L. Swatuk, C. Cash (Eds.), *Water, Energy, Food and People Across the Global South. International Political Economy Series* (p. 1–13). Palgrave Macmillan: Cham. https://doi.org/10.1007/978-3-319-64024-2_1

Tempelhoff, J. (2018). *South Africa's water governance hydraulic mission (1912–2008) in a WEF nexus context*. Durbanville: AOSIS

Tyson, P. (1993). Recent developments in the modelling of the future climate of southern Africa. *South African Journal of Science*, 89(10), 494–505

U of SA (1952). Irrigation and conservation of waters Act, No. 8, 1912. In L. Steyn, R. Mckerron, & I. Isaacs (Eds.), *The Union Statutes 1910–1947: Classified and annotated reprint* (pp.283–482). Cape Town: Peninsula Press, Limited and Government Printer

United Nations (1955). *Congress of the people, 25–26 June 1955*. Official Records of the General Assembly. United Nations Document A/AC.70/5

United Nations (1972). *United Nations conference on the human environment, 5–16, June 1972*. Stockholm: UN. Retrieved June 7, 2023, from https://tinyurl.com/26gn49sk

United Nations Environment Programme. (1987). *Report of the world commission on environment and development (Brundtland Report)*. New York: United Nations. Retrieved August 20, 2021, from https://digitallibrary.un.org/record/139811?ln=en

United Nations. (1974). *Report of the special committee against apartheid*. New York: United Nations, General Assembly. Retrieved June 19, 2023, from https://jstor.org/stable/al.sff.document.scaa1974001

Vale, P. (2023, February 10). Henry Kissinger at 100: History will judge the former US secretary of state's southern African interventions to be a failure. *The Conversation*. Retrieved April 3, 2024, from https://tinyurl.com/2me5vevd

Van Reenen, R. J. (1935). *Resisting drought*. Pretoria: Government Printer

Van Vuuren L (2013) Together we can do more – environmental consciousness in the South African dam construction sector (1945–1980). *TD: The Journal for Transdisciplinary Research in Southern Africa*, 9(1), 51–80

Vogel, C., Scott, D., Culwick, C., & Sutherland, C. (2016). Environmental problem-solving in South Africa: Harnessing creative imaginaries to address 'wicked' challenges and opportunities. *South African Geographical Journal*, 98(3), 515–530. https://doi.org/10.1080/03736245.2016.1217256

Voosen, P. (2024). 2023 was the hottest year on record—and even hotter than expected. *Science*, 383(6679), 134. https://doi.org/10.1126/science.adn9640

Watermeyer, F. (1905). Geographical notes on South Africa south of the Limpopo. *South African Journal of Science*, 1(1), 293–318. https://doi.org/10.1080/00369220608733637

Weart, S. (2003). The discovery of rapid climate change. *Physics Today*, 56(8), 30–36. https://doi.org/10.1063/1.1611350

Wessels, A. (2017). Half a century of South African 'border war' literature: A historiographical exploration. *Journal for Contemporary History*, 42(2), 24–47. https://doi.org/10.18820/24150509/JCH42.v2.2

Willems, W. (2013). Zimbabwe will never be a colony again: Changing celebratory styles and meanings of independence. *Anthropology Southern Africa*, 36(1–2), 22–33. https://doi.org/10.1080/02580144.2013.10887021

WMO (1975). *Seventh Meteorological Congress. Proceedings 28 April–23 May*. WMO. (WMO No. 428)

WMO (2023). *State of Global water resources 2022* (pp.66 1333). World Meteorological Organisation. Retrieved October 15, 2023, from https://library.wmo.int/viewer/68473/download?file=1333_en.pdf&type=pdf&navigator=1

Water Research Commission (1990). *Annual report Water Research Commission*. Pretoria: WRC

Vogel, C., Scott, D., Culwick, C., & Sutherland, C. (2016). Environmental problem-solving in South Africa: Harnessing creative imaginaries to address [illegible] challenges and opportunities. South African Geographical Journal, [illegible] https://doi.org/10.1080/03736245.2016.1217256

[illegible]

Undoing the Dualisms: Towards an Ecofeminist, Postcolonial Environmental Justice

3

Yolandi M. Coetser

Abstract

The climate crisis, driven by anthropogenic climate change, can be attributed to what I call a problematic 'view-of-nature'. This view-of-nature is an anthropocentric one that posits nature as instrumental—to be consumed and used at will. It is also historically prevalent in the Global North/West, which has also been the main driver of climate change (Povinelli, 2016). In this chapter, I briefly show how a problematic view-of-nature is salient through some of the intellectual history of the West. I argue that the triple oppression of women (through patriarchy), nature (through anthropocentrism), and the Other (through colonialism) have their roots in what Karen J. Warren calls the 'logic of domination'. This logic of domination 'posits the moral superiority of humans or groups of humans and claims that this superiority justifies the subordination of inferior groups' (Warren, 1990: 129). Thereafter, I spend more time considering how ecofeminists have provided an alternative paradigm, where dualisms, such as man/woman, human/nature and rationality/emotion are rejected in favour of a more nuanced worldview, specifically a more nuanced view-of-nature. Given that the triple oppression of women, nature and the Other stems from the same logic of domination, it follows that a postcolonial ecofeminist approach may provide a promising approach to thinking about the climate crisis.

Y. M. Coetser (✉)
North-West University, Potchefstroom, South Africa
e-mail: Yolandi.Coetser@nwu.ac.za

A. C. Terblanché-Greeff et al. (eds.), *Climate Change and the Postcolonial*, Kulturelle Figurationen: Artefakte, Praktiken, Fiktionen,
https://doi.org/10.1007/978-3-658-49328-8_3

Keywords

Ecofeminism · Climate change · Postcolonialism · Anthropocentrism · Environmental justice

Introduction[1]

It is by now established that climate change, as the world is currently experiencing it, has anthropogenic origins (IPCC, 2022). This means that contrary to the usual extended periods of planetary heating and cooling, there is now an accelerated period of planetary heating that can be directly attributed to human activity. Specifically, it is the reliance on the burning of fossil fuels over the last approximately 150 years that has led to the rapid warming of the atmosphere. The burning of fossil fuels has been especially prevalent since the Industrial Revolution, the locus of which was Europe and North America in the 18th and 19th centuries. The burning of fossil fuels catapulted the development of industrial societies. During this period, there seemed to be little awareness of the possible calamitous consequences of fossil fuel emissions, despite some of the first scientific ideas on climate change emerging during this time (Brönnimann, 2002). Nations in North America and Europe were subsequently responsible for around half of all emissions since the Industrial Revolution (Chancel, 2021), thereby reaping the benefits of the concomitant growth that resulted from the use of fossil fuels. Since Sub-Saharan Africa only produces 4% of the emissions (Chancel, 2021), it follows that, like other similar low-emission regions, it did not experience the same carbon-dependent growth during the 18th and 19th centuries than other how-emission regions did, and it still struggles to grow. This entails that these developing regions must continue to rely on fossil fuels to drive their own growth, despite knowing the harm that this may cause, or, alternatively, invest in renewable resources, the capital for which, being some of the poorest nations in the world, many of these regions may not have.[2] Additionally, it is also the case that the Global North[3] has 'greater capacity to reduce their emissions and adapt to the

[1] This chapter relies on ideas I have discussed elsewhere (Coetser, 2018).

[2] This phenomenon has been discussed at length in the literature (Barnett, 2007; Halden, 2007; Islam and Winkel, 2007; Dalby, 2013; Byrne and Maslin, 2015).

[3] I am aware of the problems with speaking broadly of the 'Global North' and the 'Global South'. This type of terminology has been accused of glossing over nuances. Although it falls outside of the scope of this chapter to delve deeper into the geopolitics around this terminological convention, I loosely follow the sentiment of Boaventura de Sousa Santas

effects of climate change' (O'Sullivan et al., 2023: 320). This is not merely a historical problem, since the carbon emissions of the Global North still far outweigh those of the Global South. By way of illustration, 'average emissions in Europe are close to 10 tonnes CO2 per person and per year', and in North America, 'the average individual emits around 20 tonnes', compared to the 1.6 tonnes in Sub-Saharan Africa (Chancel, 2021: 2).

Not only are developing nations now being pressured to reduce their own reliance on fossil fuels, but they are also the ones experiencing the effects of climate change more immediately and dramatically. For example, Pacific Island Nations are disproportionately affected by rising sea levels, despite having contributed very little to climate change (Enari and Jameson, 2021). Similarly, many regions in Africa are adversely affected by increasing droughts and floods despite negligible contributions to climate change (Zeufack et al., 2021; Maino and Emrullahu, 2022). To illustrate, according to the Global Climate Risk Index, countries such as Mozambique, Zimbabwe, the Bahamas, Puerto Rico, Myanmar and Haiti were most affected by extreme weather events at various times between 2000 and 2019 (Eckstein et al., 2021). Overall, many regions in the Global South have faced problems such as 'forced relocations, loss of property, lower quality of life and death' because of climate change (O'Sullivan et al., 2023: 321).

All of the above is 'entrenching a system of climate apartheid', rooted 'in processes of colonization, racial capitalism, and hetero-patriarchy that render some populations expendable' (Rice et al., 2022: 621). Climate justice, as a conceptual framework, allows us to interrogate these issues, There are many ways to define climate justice, but broadly, it 'recognizes the disproportionate impacts of climate change on low-income communities and communities of color around the world, the people and places least responsible for the problem' (Anon., 2025). Climate justice thus explores structural, socioeconomic and intergenerational inequalities (UNDP, 2023). Within a climate justice framework, mitigation and adaptation to climate change is neccessary to prevent further inequalities (Mary Robinson Foundation, 2011).

An important aspect of climate justice is understanding the reasons why some communities are disproportionately affected by climate change, since viewing humanity as a 'singular undifferentiated force' risks, as DeBoom argues, 'a

---

(2016: 18–19) who defines the South not geographically, but rather as 'a metaphor for the human suffering caused by capitalism and colonialism on the global level' and so can also refer to excluded, marginalised and silenced people who reside in the geographical Global North. For further discussion, see, for example, Müller (2020).

collective amnesia about the uneven historical responsibilities for climate change' (2021: 900). As such, it is worth employing another concept—i.e. postcolonial justice—to examine the 'interconnectedness and common causes of coloniality and anthropogenic climate change' (Morrison et al., 2023: 1). In this chapter, I argue that a postcolonial justice must also consider the patriarchal sentiment that underlies much of the Western view of nature, and therefore it is appropriate to consider an ecofeminist, postcolonial environmental justice in response to the environmental crisis.

The above is a mammoth, ongoing task, the intricacies of which could never be addressed in a single chapter. Nevertheless, I argue here that one can begin to construct a narrative that explains an overlap in logic that has led to the colonial project, pervasive patriarchy, and the environmental crisis. I also argue that the environmental crisis (i.e. climate change) has its origins in the same 'logic of domination', as Karen J. Warren calls it, as colonisation and patriarchy. In order to investigate this triple oppression of women, nature and the Other through patriarchy, anthropocentrism and colonialism, I first very briefly discuss how the logic of domination of nature has developed throughout some of the Western intellectual history. I do this by providing a brief historical overview of key philosophical figures, trends and epochs that have initiated, cemented and perpetuated the notion that humans are superior to nature, which also entailed that men are superior to women. In this chapter, I take my cue from Greta Gaard (2015: 20) who says, '[a] queer, posthumanist, ecological and feminist approach—brought together through the intersectional lens of ecofeminism—is needed to tackle the antifeminist threads companioning the scientific response to climate change'. I then look at how ecofeminists have provided an alternative paradigm, where dualisms, such as man/woman, human/nature and rationality/emotion, are rejected in favour of a more nuanced worldview, specifically a more nuanced view-of-nature. Given that the triple oppression of women, nature and the Other all stem from the same logic of domination, it follows that a postcolonial ecofeminist perspective may provide a promising lens through which to examine the climate crisis.

Oppression Through the Ages: A Historical Narrative of the 'Problematic View-of-Nature'

In this chapter and elsewhere (Coetser, 2018), I use the hyphenated 'view-of-nature' to denote a very specific aspect of the Global North's collective worldview. The historically prevalent view-of-nature in the Global North is one of superiority and dominion over the natural world. In this section, I very briefly discuss how the logic of domination of nature has developed throughout some of Western intellectual history. I do this by providing a brief historical overview of key philosophical figures, trends and epochs that have initiated, cemented and perpetuated the notion that humans are superior to nature. This section will, due to limited space, inevitably rely on some generalisations, however, the aim here is not to provide a comprehensive historical overview of the problematic view-of-nature. Rather, the aim is to show that there has, feasibly, been an anthropocentric trend throughout much of Western history and to punctuate this claim with key philosophical ideas. I agree with Gary Steiner (2005) who says the more we learn about our intellectual lineage, the more adeptly we can rethink our relationships with nature and animals. The dichotomy, or dualism, between humankind and the natural world has been present since antiquity, and has resulted in the anthropocentric view placing humans in a privileged position over the natural world (Newmyer, 1999).

A problematic view-of-nature can be traced back to Aristotle (384 BCE—322 BCE) who, unlike his contemporaries Homer, Hesiod and Pythagoras who saw humans and animals on a continuum, argued for a categorical difference between humans and animals (Steiner, 2005). Aristotle's distinctly anthropocentric view had a profound impact on many Christian scholars, who, in turn, had a significant impact on the Western worldview. While Aristotle thought that animals have cognitive capacities and personality traits (*The History of Animals* Book 1 Part 1), he stopped short of ascribing reason and other advanced cognitive capacities to them. Interestingly, although Aristotle has no qualms about referring to humans as 'animals'[4] (*Politics* 1253a8, *Parts of Animals*, 686a), he insists on their categorical difference (Sorabji, 1993). Animals, like slaves, are tools for man—animals provide food, labour, clothes and tools. He argues that 'since nature makes nothing purposeless or in vain, it is undeniably true that she has made all animals for the sake of man' (*Politics* 1256b).

This denial of reason, and therefore moral standing of animals has had a profound effect on the Western view-of-nature throughout the following two millennia. For example, some thinkers have lamented the fact that it was Aristotle, and not Pythagoras or Plato, who influenced the Western tradition (Singer, 2002 [1975]; Dombrowski, 1984). Even though some of Aristotle's contemporaries, like Pythagoras and Porphyry, had strong beliefs about the importance and moral

[4] In a way, this predicts the future use of 'non-human animal' and 'human-animal' that is commonplace in environmental philosophy literature. In other literature, terms such as 'other-than-human animal' (Diehm, 2006) or even 'more-than-human' (Fenske and Norkunas, 2017) are used.

status of both animals and the environment, Aristotle set the tone for some future understandings of animals and, by extension, the environment.

Aristotle's view of animals as irrational became embedded in much of Western, Latin-speaking Christendom, entailing that there remains a relative complacency in this tradition about the killing of animals (Sorabji, 1993). Many contemporary justifications for the indifference to animals and the environment can be traced back to Christian notions, which in turn were shaped by Aristotelian ideas. Some authors argue that the predominance of a Judeo-Christian view-of-nature facilitated the development of an individualism that perceives humans as apart from nature, ultimately contributing to environmental degradation (Oruka and Juma, 1994). Similarly, Lynn White Jr's controversial article, *The Historical Roots of Our Ecologic Crisis* (1967), highlights Christianity's negative impact on humanity's relationship with the natural world. He argues that Christianity is 'the most anthropocentric religion the world has seen' (White Jr, 1967: 1205).[5] The perceived separation between humanity and nature instils a 'mood of indifference to the feelings of natural objects' (White Jr, 1967: 1205). These values of indifference and separation are also found in the works of scholastics like St Augustine and St Aquinas. Both Augustine and Aquinas,[6] like Aristotle, denied animals reason and, therefore, moral status. Augustine categorically states that '[animals] are dissociated from us by their want of reason' (*City of God* 1.20). This perceived lack of reason entailed that the basic Christian tenet, 'Thou shalt not kill' should not extend to animals, since animals (and plants) were 'given' to humanity (Dombrowski, 1984; Sorabji, 1993). Despite earlier Christian thinkers expressing consideration of animals, Aquinas's reinterpretation of Christianity through Aristotle seems to have undone much of the earlier sympathy towards animals (Armstrong and Botzler, 2008).

Not all Christian thought is that of the Catholic Church, but the Protestant movement, with its emphasis on the literal interpretation of Biblical Scripture, tended to be similarly unsympathetic towards animals. The perceived divinely given dominion, in turn, alienated humans from nature. This alienation from nature, I hold, has been influential in allowing the neglect and exploitation of the environment and its non-human inhabitants.

It is difficult to deny the impact that Christianity has had on Western ways of thinking, and many scholars have discussed this at length (Whitehead, 1925;

[5] He contrasts Christianity's separation of humanity and nature with ancient paganism and Asian religions (with the exception of Zoroastrianism) (White Jr 1967: 1205).

[6] Elsewhere Callum Scott and I have considered Aquinas's animal ethics in more depth (Scott and Coetser, 2015).

Gauchet, 1997; Cloots, 2008). As André Cloots (2008) argues, Judeo-Christianity has gone along well with the rationalist tendency of the Enlightenment, which, unfortunately, perpetuated the anthropocentric trend of the Middle Ages. One thinker that exemplified this rationalist tendency is the French philosopher René Descartes (1596–1650). Rejecting the teachings of the church, he employed a more scientific approach, arguing from *a priori* premises and using observation and experimentation to discover truths about the world (Descartes, 1965 [1637]). Often touted as the 'father of modernity', he posits the natural world as a physical entity possessing fundamental properties that interact with universal laws (Hatfield, 2015). Turning his attention to animals, he explains animal behaviour in similarly mechanistic terms, and he postulated that no references to consciousness were required to explain animal behaviour (Wilson, 2010). Rather, a mechanistic explanation of being-as-machine was adequate, as it explains behaviour using simple cause-and-effect reasoning (Keeling, 1968; Allen and Trestman, 2015). Following this line of reasoning, Descartes argues that there is a categorical difference between humans and animals, evidenced by animals' lack of speech (Descartes, 1991 [1649]; Letter to More, 5 February 1649). This argument perpetuates the Aristotelian notion that animals lack fundamental capacities that categorically distinguish them from humans.

In what has been discussed thus far, one can identify a trend of anthropocentrism in some of the Western philosophical thought, which served to perpetuate the separation and alienation of humans and nature. Subsequently, the natural world was largely seen as something to be explored, exploited and dominated, and this eventually resulted in industrialisation. Industrialisation, or the mechanisation of industries including farming, played (and continues to play) a significant role in the degradation of the environment and climate change. Eric McLamb (2013) agrees that the ecological crisis can be attributed to the 'coming of age of man's ingenuity: the Industrial Revolution', which signalled a golden age of technological advancements and improvements in living conditions. Unfortunately, this golden age came at a price, as the Industrial Revolution relied heavily on fossil fuels—coal, natural gas and oil—that initially appeared to be inexhaustible. The burning of fossil fuels is arguably the single greatest contributor to the increase of greenhouse gases in the atmosphere, which, in turn, leads to global warming and climate change. The mining processes to obtain these fossil fuels are often ruthless, leading to the destruction of many natural ecosystems and destroying and reducing available habitat. The extinction of many species is also a direct result of the collection and burning of fossil fuels. One of the results of the Industrial Revolution was production on a large scale. The production that took place during the Industrial Revolution relied heavily on natural resources, and so was a specific type of natural exploitation in the guise of 'progress'. Additionally, the Age of Enlightenment, during which the Industrial Revolution took place, centred on the idea of progress and development, and subsequently venerated two categories: 'modern scientific knowledge and economic development' (Shiva 1988: xii).

A consequence of industrialisation and the focus on progress and economic development, is that it made animal and environmental exploitation a systemic problem. Yonghong Zhang concurs when they say that increasing environmental destruction synchronises with the 'capitalist industrialization and modernization process' (2013: 69). Capitalism has an inherent pursuit for profit maximisation, which, in its current state, is leading to a deteriorating natural environment (Zhang, 2013). While there had been environmental problems before capitalism, they were mostly regional with negligible impacts on the broader society, today's reality is different: 'in several hundred years of capitalist globalization and in the process of 'conquering nature' by the capitalist mode of production, the environmental problem has been becoming more and more serious and ravaging the world' (Zhang, 2013: 71). Through so many campaigns, the individual is often targeted to address climate change—stop using plastic straws, or, cycle to work, or, recycle your plastics. However, these campaigns are, in my view, a red herring—they distract from the *real* contributors to climate change, which are often multinational conglomerates. The Carbon Majors Database confirms that it is 'over half of global industrial emissions since human-induced climate change was officially recognized can be traced to just 25 corporate and state producing entities' (CDP, 2017a: 8) and that in fact, 100 active fossil fuel producers have produced 71% of all greenhouse gas emissions since 1988 (CDP, 2017b). Many of these producers form part of the primary energy sector and are fossil fuel extraction companies. Most individual energy consumers have very little choice as to where their energy comes from. Even if they can generate their own household energy through, for example, solar power, they would have very little control over the energy used by, for example, the shops they frequent, public transport, their workplace's energy consumption etc. For example, asking the average person in China to stop using plastic straws to curb climate change, when China's coal industry is the top contributor to greenhouse gas emissions, is therefore a misplaced agenda. Most modern societies' basic structures are such that it is nearly impossible to live a carbon neutral and exploitation free lifestyle. Exploitation of the natural world is hard-wired in policies, practices, habits, cultures and thoughts. Unfortunately, 'prosperity is built on fossil fuels' (Swiss Peace Foundation, 2018). Prosperity, in a capitalist reality, is strongly coupled with consumerism, which is, at least in its current form, incompatible with environmentalism. This means that as humans aim to prosper, both as individuals and collectives, they inevitably, and often unknowingly, contribute to some form of exploitation of either animal or nature. The exploitation of the natural environment becomes a mere input to attain an output—a 'cog in the machinery'. But, unless the machine is fundamentally redesigned, the cog (i.e. environmental exploitation) remains necessary, and its continued use is unavoidable. "The makings of ecological breakdown are thus inherent in capitalism", says Victor Wallis (2009: 32), and the pervasiveness and perpetuation of capitalism means that the ecological breakdown is knitted into the fabric of society.

In this section, I identified a notable trend in philosophical thought that I characterised as an anthropocentric 'view-of-nature'. This view-of-nature broadly views humans as superior to, and separated from, the natural world, and has allowed for the unabated and systematic exploitation of natural resources, which has resulted in anthropogenic climate change. In the next section, I show how this same view-of-nature has also led to the oppression of women and other nations through colonialism.

The Problematic View-of-Nature, Patriarchy and Coloniality

In this section, I argue that the triple oppression of women, nature and the Other stems from what Warren calls a 'logic of domination' (1990: 129). The logic underlying the problematic view-of-nature discussed in Section 2 has, therefore also served to oppress women and colonial subjects, since all three of these groups are 'Others' i.e. 'groups that are inferiorized, marginalized, exploited, and dominated within patriarchal and other oppressive conceptual frameworks, systems, and institutions' (Warren, 2002: 42).

Therefore, as some ecofeminists have argued, it is not anthropocentrism, but rather androcentrism that underlies a problematic view-of-nature. Put differently, it is the 'Rational Economic Man', associated with masculine traits such as egoism, rationality and calculative abilities (Plumwood, 2002) who has penetrated 'the natural world with chainsaw and bulldozer, smugly unaware of his own necessary connection with the living processes he is so intent on destroying' (Hawkins, 2009: 101). This points to an entrenched environmental patriarchy where the environment blends into the background and is made invisible against the achievements of reason and culture, provided by the white, Western, male expert (Plumwood, 1993).

One way in which this androcentrism is evident is in what Warren calls 'oppressive conceptual frameworks'. These frameworks have functioned historically to maintain, perpetuate and justify the domination of women, other subordinated humans and non-human nature—what I have referred to above as the triple oppression of women, nature and the Other. These oppressive frameworks rely on the 'oppositional value dualisms' and 'value-hierarchical thinking'—from here, power, privilege, and domination flow (Warren, 2000). These dualisms are entrenched within the Western worldview. As Yrjö Haila (2000: 160) reminds us, this opposition has historically arisen from 'the externalisation of nature in both productive and cognitive practices', especially when considering the 'culture/nature' dualism. This nature/culture opposition is, in turn, composed of many other binaries, such as woman/man, emotion/reason, nature/reason. Furthermore, dualistic thinking has had a debilitating effect on the comprehension of ecological problems, since it is nearly impossible to distinguish between 'nature as nature' and 'nature as modified by nature' (Haila, 2000: 158). Humans are part of nature, and so too is their culture, as well as the artefacts they produce as part of this culture[7] (Haila, 2000).

Dualisms are therefore not simple oppositions—rather, since one side of the binary is always seen as superior, dualisms serve to oppress. One can then see that, if nature is considered as subservient to 'man' as entrenched in the man/nature dualism, or because of the 'culture/nature' dualism, it explains the roots of anthropogenic (or androgenic) climate change. Nature has, for most of recent history, been seen as a resource to be exploited by man—and man has done so. Climate change, we now know, is caused by human activity. One might then be able to argue that if the problematic dualisms underlying the problematic view-of-nature can be shown to be false, then one can begin to weaken 'the totalizing grip of the nature-culture dualism' (Haila, 2000: 163).

The man/nature, culture/nature dualisms extend to the colonial project too.

> Ecofeminists argue that the same value systems that in Anglo-European metaphysical systems of dualism associate women with nature also have historically regarded persons of color as 'closer to nature.' These systems have constructed women and people of color as belonging to the degraded categories of corporeality, emotionality, and immanence where they are considered 'less rational' and 'less human' than the dominant group. (Mallory, 2013: 254)

Mogobe Ramose shows in 'The Struggle for Reason in Africa' (2002) how rationality, and the idea that Africans, Australasians and the indigenous populations of the Americas are not endowed with rational faculties, was

---

[7] Steven Vogel makes a similar point in *Thinking Like a Mall* (2015) where he urges his readers to think critically about the concept of 'nature'.

fundamental in colonisation and silencing these people. 'Reason, the *differentia specifica* of human beings compared with other organisms, was born from the tendency to dominate nature', argues Haila (2000). Similarly, reason was used to dominate other humans, as Ramose (2002: 2) demonstrates when he says

> Aristotle's definition of man [as rational animal] was deeply inscribed in the social ethos of those communities and societies that undertook the so-called voyages of discovery—apparently driven by innocent curiosity. But it is well known that these voyages changed into violent colonial incursions.

The Rational Economic Man therefore also penetrated the rest of the world like he did nature, which was perhaps most vividly illustrated by the Berlin Conference (1884–1885) in which European powers divided up Africa up amongst themselves, without consulting with or considering the Africans.

It should not escape one's attention that a significant characteristic of colonisation was the extraction of natural resources from the colonised regions, which further illustrates the relationship between the domination of nature and the domination of colonised 'Others'. Indeed, 'climate change makes enduring colonial injustices and structures visible' and is therefore a serious hurdle to any effort made, for example, to decolonise indigenous land (Reibold, 2023). In the literature, scholars speak of climate coloniality (Mahony and Endfield, 2018; Liboiron, 2021; Sultana, 2023) to show that '[t]he inequalities generated and exacerbated by climate change in the present have longer and connected histories once colonialism is properly acknowledged as a continuous factor' (Bhambra and Newell, 2023: 180).

Shiva (1988) argues that development, which is so often predicated on the use of fossil fuels, reduces women and nature to passive objects that are acted upon by economic and technological forces. But colonial subjects were similarly made passive objects. In South Africa, for example, traditional ways of life were destroyed, and Africans were forced to participate in capitalist labour, often in industries that contributed to the degradation of nature, like mining and farming (Carruthers, 1995). Kathryn Yusoff (2018: 14) captures the sentiment of passivity and dehumanisation in mining when she says '[h]istorically, both slaves and gold have to be material and epistemically made through the recognition and extraction of their inhuman properties'. Plumwood (1993:195) laments that 'the strands interwoven by this master story of colonisation form a mesh so strong, so finely knit and familiar it could almost pass for our own bodies, but it is an imprisoning web which encloses us'.

The logic behind colonialism and the oppression of marginalised people such as women is not only an historical cause of the oppression of people and the oppression of nature but continues systemically to affect oppressed peoples' relationship with the environment. As Kyle Whyte (2018: 125) unequivocally states '[s]ettler colonialism is a form of domination that violently disrupts human relationships with the environment'. And, if we understand settler colonialism as fundamentally androcentric, then we can consider, perhaps, what a feminist alternative would constitute.

Dualisms are entrenched in various forms of oppression, specifically gender, race and colonisation, and extend to the oppression of the natural world (Plumwood, 1993). Plumwood reiterates that the dualism that she critiques in her work pertains to a Western idea of humanity. For her, the Western model of human/nature relations (similar to what I name view-of-nature) reeks of dualism, and therefore requires anti-dualist remedies (Plumwood, 1993). Haila (2000: 156) refers to this as the challenge to develop 'concrete analyses of socioecological complexity'. A systemic change is therefore imperative, but it cannot happen, I contend, without a change in view-of-nature. A fundamental change would begin with a rejection of the dualism between humans and nature, which has led to an alienation of humanity from nature, as well as the oppression of women and colonised people. In the next section, I consider what an ecofeminist, postcolonial environmental justice would look like.

Seeking Change: An Ecofeminist, Postcolonial Environmental Justice

Chaone Mallory asks: 'How can we overcome the oppositional attitude this dualism produces, so that we can reform our scientific and political practices such that the possibilities for justice are increased?' (2013: 253). Ecofeminism can, I argue, provide a place to begin addressing this question, which Plumwood (1993: 7) affirms when she says 'there is a way of relating to the other that is especially associated with women, which contains the seeds of a different human relationship to the earth'. Mallory (2013: 253) concurs, arguing that

> struggles for environmental justice that do not incorporate an explicitly gendered and ecofeminist analysis of ecological problems will not adequately understand the ways in which systems of oppression (such as racism, colonialism, gender discrimination, and environmental degradation) are interconnected and mutually reinforcing.

Similarly, Warren (2000) argues that in order to combat the oppressive conceptual frameworks, one needs to consider feminism, local and indigenous perspectives and science, development and technology. It is therefore evident that many scholars argue that an ecofeminist account is a promising approach to addressing the environmental justice.

A dualistic way of thinking, according to Plumwood (1993), shapes the identity of both the oppressor and the oppressed, and denies their interdependence. Dualistic thinking often latches on to existing or perceived differences, and is perpetuated through systems of domination and accumulation as exhibited in social and political arrangements (Plumwood, 1993). Warren (2000) affirms Plumwood's sentiments pointing out that it is impossible to make ecologically perfect decisions or attempt an ecologically perfect lifestyle while the prevailing institutional structures are characterised by unequal wealth, consumption of energy and gendered divisions of labour. A solution therefore only begins by acknowledging the falsity of dualisms, and systemic change must follow.

Unfortunately, not much systemic change has really happened in the three decades since Plumwood published *Feminism and the Mastery of Nature* (1993), as Gaard (2015: 20) laments

> Issues that women traditionally organize around—environmental health, habitats, livelihoods—have been marginalized in debates that treat climate change as a scientific problem requiring technological and scientific solutions without substantially transforming ideologies and economies of domination, exploitation and colonialism.

One must be careful, however, when developing feminist responses to climate change do not focus on women individually, but rather '*gender as a system structuring power relations*' (Gaard, 2015: 22; Gaard's emphasis). While it is true that women, and specifically those in the Global South are more severely affected by climate change, this is not due to an innate weakness, but 'is a result of inequities produced through gendered social roles, discrimination, and poverty' (Gaard, 2015: 23).[8]

[8] While this paper is making a feminist argument, I would be remiss not to acknowledge that gender extends far beyond the binary of woman/man. In addition, men's gender roles are also affected by gender-as-a-system, as are those with non-conforming sexualities and gender identities. While important, it falls outside the scope of this chapter to discuss these issues. Nevertheless, Susie Jolly (2022) provides a poignant discussion of these issues in 'Gender, sexuality and the environment', which readers are encouraged to access if they want a more comprehensive discussion of gender and the environment.

Climate change cannot be addressed using the logic that informed the current, dominant view-of-nature—indeed, the master's tools will never dismantle the master's house. Put differently, since climate change is produced by androcentric ideology, it 'will not be solved by masculinist techno-science approaches' (Gaard, 2015: 22). Even traditional conceptions of environmental justice are limited, since they 'fail to provide promising alternative visions of global politics that can substantially combat the injustices and inequalities that are currently so popular in neoliberal environmental governance' (Mousie, 2012: 21).

Ecofeminism, especially in the form presented above, takes us towards an alternative approach, but Gaard rightly argues for an even more inclusive approach—that of a queer, feminist, posthumanist climate justice. I would argue that such an approach needs to be postcolonial too. A postcolonial ecofeminism 'interrogates how colonial/neocolonial/metropolitan interests are represented in concepts of women and nature with the ultimate aim of "decolonization' (Brazal, 2021: 221). Colonial legacies still live on in previously colonised spaces in a variety of different ways, not least of all through environmental realities. As mentioned earlier, there exists a climate apartheid, which not only entails that developing nations bear the brunt of climate change without also benefitting from the use of fossil fuels initially, but it also means that they are less able to adopt adaptation and mitigation strategies (i.e., less developed from a Western perspective). Even the IPCC (2022) report acknowledged that colonialism has not only contributed to climate change, but it also influences the way indigenous and local communities are able to respond to its current challenges.

A postcolonial orientation towards climate change then understands that justice will not be found in a dominant worldview, since it is what I termed an androcentric view-of-nature, that has led to the climate crisis in the first place. Furthermore, a postcolonial orientation then· also understands that systemic change is necessary—and this systemic change extend far beyond what is possible or doable by any individual. The onus, therefore shifts from individual responsibility to institutional, governmental, or collective responsibility of those who actually continually contribute to climate change. A postcolonial orientation also understands the shared oppressions between women, colonised people and nature, and argues for an approach that undoes the dualistic thinking that perpetuates this oppression.

Conclusion

It is impossible to make ecologically perfect decisions or attempt to live an ecologically perfect lifestyle while the prevailing institutional structures are characterised by unequal distribution of wealth, consumption of energy and gendered divisions of labour (Warren, 2000). It is for this reason that I argue for an alternative, inclusive approach. As many ecofeminists have argued, structural issues regarding race, gender and colonialisation need to be addressed as part of the fight against climate change. An alternative view-of-nature, I believe, must be informed by a postcolonial feminist perspective. This perspective provides an anti-dualist remedy against the dualism prevalent in the Western view-of-nature, which, I argue, has contributed to the alienation of humans from nature.

In this chapter, I traced this problematic 'view-of-nature' through the West's intellectual history, arguing that a dualistic and androcentric way of thinking is deeply and historically entrenched in the West. Then I considered an ecofeminist response and argued, via Gaard, that an inclusive approach needs to be taken—one that is not only feminist but also postcolonial.

References

Allen, C., & Trestman, M. (2015). *Animal consciousness*. Retrieved December 28, 2015, from https://platostanfordedu/archives/sum2015/entries/consciousness-animal/

Anon. (2025). What is climate justice? Center for Climate Justice. Retrieved April 26, 2025, from https://centerclimatejustice.universityofcalifornia.edu/what-is-climate-justice/#:~:text=Climate%20Justice%20recognizes%20the%20disproportionate,least%20responsible%20for%20the%20problem

Aquinas, T. (1955–1957). *Contra gentiles*. New York: Hanover House

Aristotle (350 BCE). *On the parts of animals*. Retrieved September 21, 2016, from https://classics.mit.edu/Aristotle/parts_animals.1.i.html

Aristotle (350 BCE). *Politics book one*. Retrieved December 10, 2014, from https://classics.mit.edu/Aristotle/politics.1.one.html

Aristotle (350 BCE). *The history of animals*. Retrieved September 23, 2016, from https://classics.mit.edu/Aristotle/history_anim.1.i.html

Armstrong, S. J., & Botzler R. G. (2008). *The animal ethics reader* (2nd ed.). New York: Routledge

Barnett, J. (2007). The geopolitics of climate change. *Geography Compass*, 1(6), 1361–75

Bhambra, G. K., & Newell, P. (2023). More than a metaphor: 'Climate colonialism' in perspective. *Global Social Challenges Journal*, 2(2), 179–187. https://doi.org/10.1332/EIEM6688

Brazal, A. M. (2021). Ethics of care in Laudato Si': A postcolonial ecofeminist critique. *Feminist Theology*, 29(3), 220–233. https://doi.org/10.1177/09667350211000614

Brönnimann, S. (2002). Picturing climate change. *Climate Research*, 22(1), 87–95

Byrne, A., & Maslin, M. (2015). Negotiating failure: understanding the geopolitics of climate change. *The Geographical Journal*,181(4), 432–436

Carruthers, J. (1995). *The Kruger National Park—A social and political history*. Durban: University of KwaZulu-Natal Press

CDP (2017a). *CDP carbon majors report 2017*. London: CDP

CDP (2017b). *New report shows just 100 companies are source of over 70% of emissions*. Retrieved May 1, 2024, from https://www.cdp.net/en/press-releases/new-report-shows-just-100-companies-are-source-of-over-70-of-emissions

Chancel, L. (2021). *Climate change and the global inequality of carbon emissions 1990–2020 Summary*. Retrieved November 21, 2023, from https://wid.world/document/climate-change-the-global-inequality-of-carbon-emissions-1990-2020-world-inequality-lab-working-paper-2021-21/

Cloots, A. (2008). Modernity and Christianity: Marcel Gauchet on the Christian Roots of the modern ways of thinking. *Milltown Studies*, 61(1), 1–30

Coetser, Y. M. (2018). *Critical re-wilding: An alternative view-of-nature*. Thesis, University of Johannesburg, Johannesburg

Dalby S. (2013). The geopolitics of climate change. *Political Geography*, 37, 38–47

de Sousa Santos, B. (2016). Epistemologies of the south and the future. *From the European South*, 1, 17–29

DeBoom, M. J. (2021). Climate necropolitics: Ecological civilization and the distributive geographies of extractive violence in the anthropocene. In *The Anthropocene*, (pp. 281–293). New York: Routledge. https://doi.org/10.1080/24694452.2020.1843995

Descartes, R. (1965). *Discourse on method optics geometry and meteorology*. Indianapolis: Bobbs-Merrill. (Original work published 1637)

Descartes, R. (1991). Letter to More 5 February 1649. *The philosophical writings of Descartes* (Volume III, pp. 360–366). Cambridge University Press. (Original work published 1649)

Diehm, C. (2006). Ethics and natural history: Levinas and other-than-human animals. *Environmental Philosophy*, 3(2), 34–43

Dombrowski, D. A. (1984). *The philosophy of vegetarianism*. Amherst: The University of Massachusetts Press

Eckstein, D., Künzel, V., & Schäfer, L. (2021). *Global climate risk index 2021*. Retrieved January 12, 2024, from https://www.germanwatch.org/sites/default/files/Global%20Climate%20Risk%20Index%202021_2.pdf

Enari, D., & Jameson, L. V. (2021). Climate justice: A pacific island perspective. *Australian Journal of Human Rights*, 27(1), 149–160. https://doi.org/10.1080/1323238X.2021.1950905

Fenske, M. & Norkunas, M. (2017). Experiencing the more-than-human world. *Narrative Culture*, 4(2), 105–110. https://doi.org/10.13110/narrcult.4.2.0105

Gaard, G. (2015). Ecofeminism and climate change. *Women's Studies International Forum*, 49, 20–33

Gauchet, M. (1997). The disenchantment of the world: *A political history of religion*. New Jersey: Princeton University Press

Haila, Y. (2000). Beyond the nature-culture dualism. *Biology & Philosophy*, 15, 155–175

Halden, P. (2007). The geopolitics of climate change: Challenges to the international system. Sweden: Swedish Defence Research Agency

Hatfield, G. (2015). *René Descartes*. Retrieved January 26, 2016, from https://plato.stanford.edu/entries/descartes/

Hawkins, R. (2009). Extending Plumwood's critique of rationalism through imagery and metaphor. *Ethics and the Environment*, 14(2), 99–113

IPCC (2022). Summary for policymakers. In H.-O. Pörtner, D. C. Roberts, M. Tignor, E. S. Poloczanska, K. Mintenbeck, A. Alegría, M. Craig, S. Langsdorf, S. Löschke, V. Möller, A. Okem, & B. Rama (Eds.). *Climate change 2022: Impacts, adaptation, and vulnerability. contribution of working Group II to the sixth assessment report of the intergovernmental panel on climate change*. Cambridge University Press, Cambridge, UK and New York, NY, USA, pp. 3–33. https://doi.org/10.1017/9781009325844.001

Jolly, S. (2022). Gender sexuality and the environment. In K. Sims, N. Banks, S. Engel, P. Hodge, J. Makuwira, N. Nakamura, J. Rigg, A. Salamanca, & P. Yeophantong (Eds.) *The Routledge Handbook of global development* (1st ed.). New York: Routledge

Liboiron, M. (2021). *Pollution is colonialism*. Durham, NC: Duke University Press. https://doi.org/10.1215/9781478021445

Mahony, M., & Endfield, G. (2018). Climate and colonialism, WIREs. *Climate Change*, 9(2), e510

Maino, R. & Emrullahu, D. (2022). Climate change in Sub-Saharan Africa fragile states. *Evidence from Panel Estimations*. Retrieved July 28,2025 from https://www.imf.org/-/media/Files/Publications/WP/2022/English/wpiea2022054-print-pdf.ashx

Mallory, C. (2013). Environmental justice, ecofeminism, and power. In R. Rozzi, S. T. A. Pickett, C. Palmer, & J. B. Callicott (Eds.) *Linking ecology and ethics for a changing world: Values, philosophy, and action* (pp. 251–258).

Mary Robinson Foundation (2011). *The geography of climate justice*. Retrieved February 27, 2025, from https://www.mrfcj.org/pdf/Geography_of_Climate_Justice_Introductory_Resource.pdf

McLamb, E. (2013). *The continuing ecological impact of the industrial revolution*. Retrieved January 4, 2016, from https://ecologyprime.com/the-ecological-impact-of-the-industrial-revolution/

Morrison, K., Nand, M. M., Ali, T., & Mele, S. (2023) Postcolonial lessons and migration from climate change: ongoing injustice and hope. *npj Climate Action*, 2(22), 1–10. https://doi.org/10.1038/s44168-023-00060-7

Mousie, J. (2012). Global environmental justice and postcolonial critique. *Environmental Philosophy*, 9(2), 21–46

Müller, M. (2020). In search of the Global East: Thinking between North and South. *Geopolitics*, 25(3), 734–755. https://doi.org/10.1080/14650045.2018.1477757

Newmyer, S. T. (1999). Speaking of beasts: The Stoics and Plutarch on animal reason and the modern case against animals. *Quaderni Urbinati di Cultura Classica*, 63(3), 99–110

O'Sullivan, A., Omukuti, J., & Ryder, S. S. (2023). Global surpluses of extraction and slow climate violence: A sociological framework. *Sociological Inquiry*, 93(2), 93, 320–340. https://doi.org/10.1111/soin.12518

Oruka, H. O., & Juma, C., (1994). Ecophilosophy and parental earth ethics. In Oruka, H. O. (Ed.) *Philosophy humanity and ecology ACTS press* (pp. 115–129). Nairobi

Plumwood, V. (1993). *Feminism and the mystery of nature*. London: Routledge

Plumwood, V. (2002). *Environmental culture: The ecological crisis of reason*. London: Routledge

Povinelli, E. (2016). *Geoontologies: A requiem to late liberalism*. Durham, NC: Duke University Press

Ramose, M. (2002). The struggle for reason in Africa.

Reibold, K. (2023). Settler colonialism, decolonisation, and climate change. *Journal of Applied Philosophy*, 40(4), 624–641. https://doi.org/10.1111/japp.12573

Rice, L., Long, J. & Levenda, A. (2022). Against climate apartheid: Confronting the persistent legacies of expendability for climate justice. *Environment and Planning E: Nature and Space*, 5(2), 625–645. https://doi.org/10.1177/2514848621999286

Shiva, V. (1988). *Staying alive: Women, ecology and survival in India*. New Delhi: Indraprastha Press

Sultana, F. (2023). The unbearable heaviness of climate coloniality. *Political geography*, 99, 102638. https://doi.org/10.1016/j.polgeo.2022.102638

Scott, C. D. & Coetser, Y. M. (2015). Rewriting Aquinas' animal ethics: The primacy of reason in the determination of moral status. *South African Journal of Philosophy*, 34(3), 289–303

Singer, P. (2002). *Animal Liberation*. New York: HarperCollins Publishers. (Original work published 1975)

Sorabji, R. (1993). *Animal minds and human morals: The origins of the Western debate*. London: Duckworth

Saint Augustine Bishop of Hippo (426). *The city of God*. Retrieved October 12, 2016, from https://www.gutenberg.org/cache/epub/45304/pg45304-images.html

Steiner, G. (2005). *Anthropocentrism and its discontents: The moral status of animals in the history of Western philosophy*. Pittsburgh: University of Pittsburgh Press

Swiss Peace Foundation (2018). *Systemic change to respond to environmental conflicts*. Retrieved April 28, 2024, from https://reliefweb.int/report/world/systemic-change-respond-environmental-conflicts

UNDP. (2023) *Climate change is a matter of justice—here's why*. Retrieved February 27, 2025, from https://climatepromise.undp.org/news-and-stories/climate-change-matter-justice-heres-why

Vogel, S. (2015). *Thinking like a mall: Environmental philosophy after the end of nature*. Cambridge, MA: MIT Press

Wallis, V. (2009). Beyond "green capitalism". *Monthly Review*, 61(9), 32–48

Warren, K. J. (1990). The power and promise of ecological feminism. *Environmental Ethics*, 12(2), 125–146

Warren, K. J. (2000). *Ecofeminist philosophy: A Western perspective on what it is and why it matters*. Oxford: Rowman & Littlefield

Warren, K. J. (2002). Response to my critics. *Ethics and the environment*, 7(2), 39–59

White, Jr. L. (1967). *The historical roots of our ecologic crisis science*, 155(3767), 1203–1207

Whitehead, A. N. (1925). *Science and the modern world*. New York: The Free Press

Whyte, K. (2018). Settler colonialism, ecology, and environmental injustice. *Environment and Society*, 9, 125–144. https://doi.org/10.3167/ares.2018.090109

Wilson, S. D. (2010). *Animals and ethics*. Retrieved December 22, 2015, from https://iep.utm.edu/animals-and-ethics/

Yusoff, K. (2018). *A billion black Anthropocenes or none*. Minneapolis: University of Minnesota Press. https://doi.org/10.1093/isle/isz077

Zeufack, A. G., Calderon, C., Kambou, G., Dijofack, C. Z., Kubota, M., Korman, V., & Cantu Canales, M.C. (2021). *An analysis of issues shaping Africa's economic future*. Washington DC: World Bank. Retrieved July 28, 2025 from https://openknowledge.worldbank.org/bitstreams/a5ef8ce3-cfdf-5187-a8ff-1162787fb1ef/download

Zhang, Y. (2013). Capitalism and ecological crisis. *Journal of Sustainable Society*, 2(3), 69–73

Postcolonial and/or Decolonial Perceptions in Framing Different Responses to the Climate Crisis

4

Michelle Pressend

Abstract

The dominant responses to the climate crisis are heavily reliant on planetary management discourses, techno-scientific efficiency approaches and market-based mechanisms. A key assumption of climate science is that climate change can be addressed by more energy-efficient technology to reduce emissions and, at the same time, catalyse economic growth. Present geo-political power struggles over technological advancements and financing in the international climate change negotiations arena situate the age-old political economy state power struggles. These struggles position the Global North as leaders in shaping technological advancements, while the Global South demands financing to address the historical colonial debt. This financing is required not only to adapt to the consequences of climate change but also to support the Global South's need for 'catch-up' development. However, despite the good intentions behind climate mitigation and adaptation, the colonial hierarchical and racial social relations that made the plantations and the coal plants in the first place, together with the expanding global neoliberalism, remain intact. This chapter illustrates that post-colonialism critically engages with imperial oppression, and post-colonial states in the Global South demand redress for climate change. However, the key to addressing racial and colonial exclusion is through advocating for inclusion and access to Western modernity's political, cultural,

M. Pressend (✉)
University of Cape Town, Cape Town, South Africa
e-mail: Michelle.Pressend@uct.ac.za

A. C. Terblanché-Greeff et al. (eds.), *Climate Change and the Postcolonial*, Kulturelle Figurationen: Artefakte, Praktiken, Fiktionen,
https://doi.org/10.1007/978-3-658-49328-8_4

economic and social praxis of progress and human 'mastery' over nature. I argue that relational and decolonial politics emanating from the knowledge of the peoples that are being erased, marginalised and rendered invisible are needed to create openings for pluriversality (Escobar, 2018) that support life-affirming solutions for addressing the climate crisis.

Keywords

Catch-up development · Just transition · Green technologies · Relationality · Pluriverse

Introduction

The mainstream anthropocentric responses to the climate crisis assume that variants of progress such as green energy and technology development within a 'win-win' framework of sustainable development will mitigate climate change (IPCC, 2007; Steffen et al., 2007; Crutzen and Schwägerl, 2011; United Nations Sustainable Energy for All, 2011). In the grand narrative of the Anthropocene, the rise of fossil fuels in the Industrial Revolution was crucial for the beginning of the Anthropocene and coincided with the invention of the steam engine in 1784 (Crutzen and Schwägerl, 2011). The accelerated burning of fossil fuels since the Western Industrial Revolution (c. 1760–1840) has caused the Earth's temperature to increase drastically. The climate crisis is understood as a consequence of human-induced carbon dioxide and other greenhouse gas emissions into the atmosphere. This reasoning is the basis of the climate science consensus (Bonneuil and Fressoz, 2016). Furthermore, the use of nuclear weapons in World War II and the Great Acceleration of the 1950s are also regarded as significant developmental periods in the Anthropocene (Bonneuil and Fressoz, 2016). The concept of the Anthropocene is largely accepted as the cause of the acceleration of global warming and the unprecedented rate of increase of the Earth's atmospheric temperature. Isabelle Stengers (2015) suggests that the term 'Anthropocene' should refer particularly to the consequences of the anthropogenic behaviour of Western civilisation and not be understood as a universal term implying that all civilisations in this geological epoch are equally culpable for climate change. Similarly, Baskin (2015: 15) questions the universal nature of human culpability in the Anthropocene discourse and observes that the concept of the Anthropocene does not distinguish 'between the different societies, both spatially and temporally', nor does it question 'which humans exactly were agents'. Baskin (2015: 16) further notes that the Anthropocene may reveal the power of humans and technology, but it also 'conceals who and what is powerful and how power is concealed'.

In relation to modern energy generation, the shift to 'cheap' coal and later oil and gas fundamentally shaped the 'energy metabolism' that in turn shapes cities, manufacturing, transport, etc., globally (Chakrabarty, 2009; Mitchell, 2009). Chakrabarty (2009: 208) writes, '[t]he mansion of modern freedom stands on the ever-expanding base of fossil-fuel use. Most of our freedoms so far have been energy intensive'. When Chakrabarty speaks about 'our freedoms' in the Anthropocene discourse, he refers to the universal human. But while modern advances in technology may have brought conveniences to many humans, what is often overlooked is that the productivist system is a constitutive relationship that is modernity's progress, which manifests as economic prosperity, well-being and social stability for some humans at the expense of others and of non-human nature (Moore, 2015b). Even when post-colonial scholars call for bridging the nature/human divide (Chakrabarty, 2009) and see humans as part of nature, nature is reduced to an output (Moore, 2014). Moore states:

> Even when our philosophical position regards humans as part of nature, the narrative rules, methodological premises, and theoretical frames of world-historical scholars often remain within the confines of a modernist view of nature as external. (Moore, 2014: 289)

Critical scholars undertheorise 'how nature matters to capitalism not merely as output, but as constitutive relations' (Moore, 2014: 289). The monetary gain from extraction means that someone or something is exploited elsewhere, so only a few benefit financially and materially at the expense of others. This is because Western modernity's objectifying epistemology, which is reductionist, rationalist and rooted in Cartesian dualism, glosses over many inconvenient facts that capital accumulation thrives at the expense of others and nature. Capitalism as world-ecology intertwined with power, profit and life, relates to values of domination and appropriation of both humans (women, slaves, indigenous and black/brown people) and non-human nature, such as forests, soils or rivers, for capital accumulation (Moore, 2014, 2015a, 2015b, 2016). The history of modern science embedded in the core notions of rationality and efficiency continues to see 'other' ways of being and knowledge as primitive and inferior (Crist, 2016). Modernity's ontology of separation, as Patel and Moore argue (2020: 202), has not only 'worked to describe and categorize the world but *served practically* to dominate and cheapen the lives of nearly all humans and the rest of nature'. The Anthropocene discourse undermines alternative human relations and 'our' place in nature and the web of life (Mignolo, 2017; Crist, 2016). Crist (2016) points out that in the Anthropocene, the 'other' (indigenous people, people of colour, women, and nonhuman worlds) has been *de facto* silenced.

This snapshot brings critical perspectives of the Anthropocene discourse to the fore and lays the foundation for this chapter. It illuminates the limitations and blind spots of post-colonial states, highlights the so-called developing countries' demands for redress in climate action, and emphasises the need for inclusive politics and 'catch-up' development. As Nabudere (2011: 15) stresses, 'We cannot [...] maintain a viable, sustainable economy by merely mitigating or adjusting to the consequences of the industrial devastation while at the same time maintaining the exploitative systems'. This chapter illustrates that post-colonialism critically engages with the dimensions of imperial oppression in the 'Global South', demanding redress for climate change. However, the key here is to address racial and colonial exclusion, characterised by techno-scientific fixes within the singular story of development, progress, and growth. Thus, the demands are primarily about inclusion and access to Western modernity's political, cultural, economic and social praxis of progress and development 'catch-up' (Mies, 2014). I argue that relational and decolonial politics rooted in the knowledge of the peoples of the Global South who are being erased, marginalised, and rendered invisible are needed to create openings for pluriversality (Escobar, 2018) in addressing the climate crisis.

First, I discuss the importance of unpacking the stories of history as a modernist project that has its roots in the dualist logic of separating humans' entanglements with nature. Second, I address the way dualist logic underpins the context of global science-based policy-making that has shaped mainstream responses to climate change. Third, I reflect on how statecraft and the concentration of power make territories and the biosphere accessible, legible, knowable, and utilisable for techno-scientific advancements and capital accumulation. Fourth, I discuss the limitations of just transitions and energy democracy demands in the Anthropocene discourse. Lastly, given these limitations and blind spots in the dominant climate change narratives, I illustrate the need for ontological shifts and the need to assess what can be learnt from non-dualist relationships that are orientated towards restoring life-affirming ways of being in practice (Salleh, 1997; Escobar, 2011; Nabudere, 2011; Mignolo, 2016, 2017; Patel and Moore, 2020). Regenerative and reciprocal praxis is about the production and reproduction of life and the well-being of all; it is the opposite of Western modernity's growth. Indigenous cosmologies and knowledges embody these relational values. As Ariel Salleh (1997: 159) notes, 'most indigenous men and women already share these links: are they willing to show us how?'.

Unpacking the Stories of History as a Modernist Project in the Anthropocene Discourse

Mainstream literature and international climate reports on human impacts on the environment cover aspects of population growth, green economic growth, market incentives for a low carbon economy and 'green' infrastructure and technology to name a few (Goodland and Daly, 1996; Ehrlich, 1971; IPCC, 2007; Steffen et al., 2007; Crutzen and Schwägerl, 2011; UNEP, 2011). According to Crist (2016), Anthropocene discourse chiefly focuses on the population increase projections leading to increased energy demands, economic growth and consumer culture as leading social models set to ensure that everyone can be affluent, embracing a positive attitude towards the prospects of human development on the planet, the belief that techno-scientific technology fixes will solve the climate crisis and developing pathways that imbue a planetary management mindset to managing the Earth's natural systems. Although the Anthropocene discourse has been problematised by several prominent scholars such as Donna Haraway, Anna Tsing, Isabelle Stengers, the late Bruno Latour, Jason Moore and Eileen Crist, among others, the Anthropocene remains the central term/concept in explaining global warming. As Moore puts it, 'the problem is not the "Age of Humans" but the "Age of Capital", not Anthropocene, but *Capitalocene*' (Moore, 2015a: 2). He emphasises that the Anthropocene describes *what* the cause of global warming is but does not describe *how* it came about (Moore, 2015a, 2015b).

In this section, I attempt to unpack some of the broad ahistorical blind spots of the climate science that shape the current policy imperatives. I do so by engaging and interrogating the normativity of Western modernity's ontology that has shaped the Anthropocene discourse and 'green thought'. Western ontology is one rooted in dominance, hierarchies, separation, violence and extraction (Tsing, 2012; Moore, 2015b; Haraway, 2016; Escobar, 2018; Davis *et.al.*, 2019; Patel and Moore, 2020; Ladha and Murphy, 2022). As Moore (2015b: 40) explains, 'green thought' has embraced an environmental perspective that emphasises the 'environmental history *of* social relations (Nature-plus-Society), rather than modernity's social relations *as* producers and products of the web of life (society-in-nature/nature-in-society)'. Even those that are critical of the impacts of capitalism on nature from the perspective of 'green' thought, focus on the consequences of 'nature-as-resource', which sees nature as outputs (Moore, 2014, 2015a; Crist, 2016; Szeman and Wenzel, 2021).

Within the mainstream discourse, post-colonial thought on the mitigation of climate change elicits the 'productivist model' (Moore, 2015a) and 'resource logic' (Howe, 2019; Szeman and Wenzel, 2021). In the so-called geopolitical Global South, governments are immersed in development 'catch-up' within the singular story of development, progress, and growth (Mies, 2014). In this way of thinking, the so-called 'developing' countries need to prioritise their industrialisation rather than industrialised centres of the world taking steps to limit their own consumption and production (Mies, 2014). Little attention is paid to framing the climate crisis beyond the realms of economic production and endless human use (Szeman and Wenzel, 2021). The singular one-world story of a universal history of development as progress remains at the centre of the Anthropocene discourse. As Szeman and Wenzel (2021: 509) stress, nature is simply seen as available 'stuff waiting to be sold and used under the logic of capitalism (and socialism, for that matter)'. Other relationships and values, including ecological, spiritual, cultural, or aesthetic, are overlooked in the Anthropocene discourse (Salleh, 2010; Crist, 2016; de la Cadena(2019); Szeman and Wenzel, 2021).

The advocacy efforts are concerned with what humanity does to Nature through capitalism, thus efforts are focused on planetary management 'solutions' (WCED, 1987). Unseen in 'green thought' is the essence of constitutive social relations rooted in human mastery over nature, in which the value of 'Cheap Nature' (both human and non-human nature) is rendered exploitable and where some humans continue to be 'othered' in 'green thought'. Moore also writes that:

> Green Arithmetic, in other words, offers a Human/Nature binary that can proceed only by converting the living, multi-species connections of humanity-in-nature and the web of life into dead abstractions – abstractions that connect to each other as cascades of consequences rather than constitutive relations. (Moore, 2017: 598)

The constitutive relations from a world history of dominance, violence, alienation, and separation made it possible to launch and sustain the process that '[put] the whole of nature to work for capital', according to Moore (2015a: 3). He explains that the 'job of science' made nature an object of capital accumulation and transformed it into extractable units, the 'job of the economy' was to channel this alienation and separation through the cash nexus and the 'job of the state' was to enforce the cash nexus (Moore, 2015a). He adds, 'it was a world-praxis of remaking the world in the image of capital—or should we say, in accord with the fantasies of capital' (Moore, 2015a: 3).

The remaking of the world by the colonial Western modernity's ontology has its roots in the 15th and 16th centuries in Europe when the understanding of the human relationship with the Earth changed from one of a holistic, organic cosmos to a mechanistic and exploitative relationship driven by 'progress' (Merchant,

1980; Federici, 2018). The decline of feudalism in Europe led to internal and external colonialism and the rise of capitalism (Moore, 2002). Moore (2015a, 2015b) describes three historical processes fundamental to the rise of capitalism. The first process was human activity transformed into the labour power cash-nexus; in the second, land was conceived of as property; and in the third, a symbolic-knowledge regime was premised on separation—on *alienation* of the fundamental interdependence of everything that exists (Moore, 2015a, 2015b).

At that time, the beginnings of the European land enclosures caused a widespread shift from subsistence to commodity production (Moore, 2002; Nabudere, 2011; Tsing, 2012).[1] As Moore puts it,

> Cheapness as world-historical strategy joined imperial power, racism, patriarchy, and accumulation to enable the greatest environment-making revolution since the dawn of settled agriculture some 8000 years earlier. This environment-making revolution pivoted on the ways that the ruling abstractions of Civilization and Savagery, or what today we call Society and Nature, joined with financialized imperialism to create, and expand the world proletariat, dramatically in the two centuries after 1550. (Moore, 2021: 161)

According to Patel and Moore (2020: 202), modernity's ontology of separation has not only 'described and categorized the world but *served practically* to dominate and cheapen the lives of nearly all humans and the rest of nature'. Conquest, productivity, and plunder became common sense and commonplace (Patel and Moore, 2020). The society/nature binary reflects real abstractions that both describe the world and make it (Patel and Moore, 2020). These abstractions or generalisations are statements of ontology and epistemology; therefore, they are invisible, and their violence is hidden (Patel and Moore, 2020).

A key part of modernity's process privileges substance and matter as things that can be isolated rather than seeing the relationships between substance and

[1] Land enclosures in Europe were a key element of the agricultural revolution and the beginning of transforming land into landed property. According to Nabudere (2011: 29), 'since land (or earth) was not the product of labour, it did not have money "value". It was sold or "rented" by the landlord as landed property to receive annual revenue and became "fictitious capital" and was treated as a pure financial asset which was bought or sold according to the rent and price it yielded'. Moore (2002: 312), points out that 'Western Europe landlords responded to the agrarian crisis' during the feudal period by 'enclosing common lands and shifting from arable to animal husbandry, especially sheep raising'. Furthermore, according to Anna Tsing (2012), the class hierarchical order was intensified by agricultural production, by which the nation state secured private property and constructed social class lines between communities, families and even individuals.

matter (Merchant, 1980; Cronon, 1983; Moore, 2015a, 2015b; Patel and Moore, 2020; Szeman and Wenzel, 2021). William Cronon's book *Changes in the Land* (1983) describes changes to the New England landscape between the 1600s and 1800s, at the time of European settlement. Cronon (1983: 21) asserts that on arrival, the colonists saw a land of plenty and sought 'merchantable commodities', but 'seeing landscapes in terms of commodities meant something else as well: it treated members of an ecosystem as isolated extractable units'. Mignolo (2011: 3) points out that 'modernity is a complex narrative whose point of origin is Europe'; this narrative is constructed on 'celebrating the achievements of Western civilisation while hiding its darker side, that of coloniality'.

The concept coloniality was coined by Peruvian sociologist Quijano (2007), who explains that human exceptionalism is ongoing, privileging 'whiteness' and Western hegemonic discourse in global power at present (Mignolo 2011, 2016). The colonial nature of being continues to be 'invasive, ongoing and invisible to settler societies' (French et al., 2020: 2).

Blind Spots of Global Climate Science Shaping Policy Design

These perspectives on modernity's human/nature relations in responses to the climate crisis are critical for looking beyond utilitarian human-centric discourse dominating the climate change policy arena. These scholarly works bring fundamental questions about a world dominated by Western modernity's views and ongoing capitalist extractive relations.

The creation of the Intergovernmental Panel on Climate Change (IPCC) established by the World Meteorological Organisation and the United Nations Environment Programme in 1988, was instrumental in 'provid[ing] governments at all levels with scientific information that they can use to develop climate policies' (IPCC, n.d.). Fioramonti (2014: 69) writes that 'scientists from all over the world began working on collating information with the view of generating consensus on the state of the planet's climate', their main goal being 'to review piles of numbers and clarify whether the globe was warming or not'. While scientists worked on statistics, economist introduced 'models to estimate the pros and cons of climate change-inspired reforms' (Fioramonti, 2014: 69). As Crist (2016: 25) notes, the 'focus on relentless overlay of data, usually numbers' collated 'into the pure truth' that supports the development of a cohesive discourse of reducing carbon dioxide and other greenhouse gas emissions through technology and market-centred approaches. These solutions in the dominant environmental and 'green' discourses prescribe technology fixes to address climate change through the advancement of 'green' technology approaches, such as geoengineering and transitions to renewable energy technologies (Crutzen and Schwägerl, 2011; United Nations Sustainable Energy for All, 2011).

Globally, those that support the idea of 'green' economic growth regard renewable energy technologies as a lucrative investment opportunity to stimulate jobs and improve the well-being of people and the planet, especially in the 'developing' countries (International Energy Agency, 2017; IPCC, 2022). However, as Crist (2016) points out, these policy imperatives are centred on progress and the development of human enterprise in the hope that the way forward will materialise in the form of green energy and economic development for all. Escobar (1996) warned that these developments are largely centred on the belief that economic growth and human development trump all living and non-living beings on the planet. Furthermore, policy imperatives try to reconcile economic growth and environment preservation without any adjustment to the market system (Escobar, 1996).

The Kyoto Protocol was adopted in 1997 and ratified in 2005 to operationalise the United Nations Framework Convention on Climate Change (UNFCCC) by committing industrialised countries to reduce their greenhouse gases with agreed targets. A range of market-based schemes to encourage industries to shift to more efficient technology to reduce pollution was introduced in the Kyoto Protocol. The idea of emission trading—popularly known as carbon trading—was introduced. The market-driven emission-trading scheme permits and credits intended to encourage fossil fuel-intensive industries to reduce and offset their emissions (Dalsgaard, 2013). A core part of the 1997 Kyoto Protocol was offering countries flexibility in meeting their emission targets as compensation for the costs incurred. To reductionist thinkers, these market-based instruments offered a 'win-win' opportunity for energy transition that would encourage high-use fossil fuel corporations to reduce their emissions while continuing to make profits (Dalsgaard, 2013). Under this trading scheme, 'the government puts a cap or limit on which each industrial actor is allowed to emit, measured in carbon permits' (Dalsgaard, 2013: 84). The cap-and-trade mechanism allows companies that exceed their limit to purchase additional carbon credits from those with surplus credits. Economic rationality wholeheartedly supports the role of market-based initiatives, and studies suggest that more effective regulatory frameworks from the state are necessary for market-based mechanisms (Pollitt, 2012; Fay, 2013; Finon, 2013). However, the reductionist approach and market-based initiatives to promote low-carbon energy remain a conundrum, as industries that buy credits do not reduce their emissions below the cap (Dalsgaard, 2013). A more fundamental concern stressed by Steffen Dalsgaard (2013) is whether any 'real' carbon reductions or 'offset' actually take place when climate mitigation is dependent on the actions and values of finance traders?

In 2015, governments from around the world committed to the Paris Agreement, adopted in 2015 at COP 21, with the goal to limit global warming to well below 2 °C, preferably to 1.5 °C, compared to pre-industrial levels (UNFCCC, n.d.). The Paris Agreement is considered a landmark in the multilateral climate change process, stating that 'for the first time, a binding agreement brings all nations into a common cause to undertake ambitious efforts to combat climate change and adapt to its effects' (UNFCCC, n.d.). However, the principle of 'common but differentiated responsibility' (CBDR), the cornerstone of the Kyoto Protocol and committed developed countries to legally binding targets to reduce emissions, was diluted in the Paris Agreement (Khare, 2016). While it is critical for countries highly dependent on fossil-fuel energy for their economic activity like South Africa to consider alternatives to current energy sources, there has been a 'shift away from the binary approach to differentiation that was at the heart of the Kyoto Protocol, towards a more flexible approach that encompasses all countries' (Khare, 2016: 109), especially the so-called emerging countries. The COP21 negotiations created commitments for 'developing' countries to set emission reduction targets (Bidwai, 2015). The Paris Agreement marked the turning point 'toward a more flexible approach that encompasses all the countries' (Khare, 2016: 109), but it does not mention inequities and historical responsibilities. The CBDR principle is mentioned in the Paris Agreement, but Rishika Khare (2016: 109) writes that 'it is more of an expression than a principle commanding strict adherence'. She infers that the wording in the Paris Agreement effectively removed the essence of the CBDR principle:

> The Parties to this agreement in pursuit of the objective of the Convention, and being guided by its principles, including the principle of equity and common but differentiated responsibilities and respective capabilities, in the light of different national circumstances. (Khare, 2016: 109)

'In the light of different national circumstances' refers to a country's capability, usually technical. For 'developing' countries considered to have limited technical capability, the underlying suggestion is that developed countries support green technological leapfrogging and investment. The dominant discourse is imagined and articulated in what Howe (2014: 383) terms an *anthropocentric ecoauthority* 'predicated on a series of experimental, scientific, and managerial truth claims regarding ecological and future forecasting in an era of global anthropogenic change'.

The nature of the financial assistance that supports greener and cleaner technology in 'developing' countries increasingly encourages public investment to leverage funds directly through private finance (Bond, 2012). When South Africa hosted COP 17 in Durban in 2011, a major outcome was the establishment of the Green Climate Fund (GCF) and the dissolution of previous funding vehicles such as the Adaptation Fund. The purpose of the GCF is to support the climate mitigation and adaptation efforts of developing countries:

> Our innovation is to use public investment to stimulate private finance, unlocking the power of climate-friendly investment for low-emission, climate-resilient development. To achieve maximum impact, GCF seeks to catalyse funds, multiplying the effect of its initial financing by opening markets to new investments. (Green Climate Fund, n.d.)

Daniela Gabor (2021) warns that the creation of favourable conditions for private sector investment means that severely indebted countries would need to facilitate the financial de-risking of the investor. Moreover, she states, 'the inclusion of institutional investors, from hedge and pension funds to insurance companies and sovereign wealth funds, and asset managers as critical stakeholders, upgraded the de-risking renewables strategy into a full-blown, ambitious "development as de-risking" paradigm' (Gabor, 2021: 434). In this way, the risk is transferred to the state's balance sheet (Gabor, 2021). Development as a de-risking paradigm puts 'states under pressure to institutionally codify risk-proofing arrangements, guaranteeing private financial profits in the name of aligning sustainable projects with the preferred risk/return profile of institutional investors' (Gabor, 2021: 453).

Gabor (2021) emphasises the financial engineering of de-risking—the state extends the infrastructural power of finance from monetary and fiscal policy to other areas of public policy. This way, the scope for a 'green' developmental state is narrowed, and industrial policy amounts to little more than planning and overseeing public-private partnership projects. She warns that the de-risking by accumulation paradigm has detrimental implications by placing the burden on the 'poor' in so-called developing countries, narrowing the policy scope of 'developing' countries' states.

Limitations of the Call for Energy Democracy and Just Transitions

The emerging politics of renewable energy transitions have catalysed demands for energy democracy and just transitions (Swilling and Annecke, 2012; Stephen and Burkes, 2018; Sweeney and Treat, 2018; Galgóczi, 2020). Non-governmental organisations (NGOs) and grassroots social movements call for energy democracy in renewable energy transitions, particularly considering concerns that renewable energy development seems to be echoing historical conflicts around the development of technologies such as hydroelectric and nuclear power (Burke and Stephens, 2018: 78). Burke and Stephens (2018: 78) suggest energy democracy can be understood as a contemporary expression of decentralised grassroots movements from the 1970s and 1980s that called for direct local action and visions of 'technology democracy'.

They point out that energy democracy discourse has its roots in various activist initiatives in Europe and the United States, such as anti-nuclear activism and concerns about the geopolitical instability of fossil fuels (Burke and Stephens, 2018). They also suggest that 'energy democracy advocates recognize that energy systems are inseparable from larger social and economic patterns and relationships and therefore energy democracy requires a careful, inclusive and strategic construction of alliances' (Burke and Stephens, 2018: 79). They note that while energy democracy does not have a specific definition, 'it is part of a process of ongoing struggles for economic and political democratization as expressed through the practical project of energy transitions' (Burke and Stephens, 2018: 79). Social movements working to address climate and economic crises are not only resisting fossil fuel use and the market-driven green economy agenda, but are also advocating for decentralised, democratised, and community-based energy futures (Burke and Stephens, 2018).

The concept of just transition originated from trade unions in the Global North in relation to compensation and social protection as a consequence of imminent job losses in energy transitions (Sweeney and Treat, 2018; Galgóczi, 2020). Since jobs and the workforce are often mentioned as the biggest 'losers', the just transition discourse has expanded to ensure that 'green jobs' are considered a 'win-win' for labour and industries (Stevis and Felli, 2015). As is the case with energy democracy, there is no single understanding of just transition. A key framing is transition from exploitative and extractive systems to regenerative, inclusive economies that address historical harm while creating equitable and just outcomes (Climate Justice Alliance, n.d.).

Concerns are emerging about the co-optation of just transition by transnational corporations and hegemonic countries to foster green growth (Stevis and Felli, 2015). Energy democracy and transition are fundamentally political (Burke and Stephens, 2018). Energy transition cannot be based solely on replacing technology—'it cannot avoid re-ordering social and political relations', and Burke and Stephens (2018: 80) propose 'deep structural shifts in energy systems as well as socioeconomic political systems'. Burke and Stephens (2018) call for an energy democracy agenda that seeks to advance democratisation and participation through democratically planned and publicly and community-owned and operated renewable energy systems that serve the public interest. Energy democracy further calls for tangible community benefits, such as decent and stable employment, public space, and transport and new public institutions. They argue that 'energy democracy seeks to empower low-income communities and communities of color' (Burke and Stephens, 2018: 79), embracing the idea that those most marginalised are well-positioned to envision and lead alternative energy futures. Just transitions must tackle structural inequalities and foster approaches that change consumption in terms of socio-economic inequalities and restoration, reconstruction, and redistributive justice (Swilling and Annecke, 2012). Implicit in this understanding is the idea that new technology presents an opportunity for deeper engagement and 're-imagining the construction of large-scale *infrastructural technologies and investments'* (Burke and Stephens, 2018: 79). In this way, the form of renewable energy policies can influence the possibility of more democratic futures, which can be achieved by building alliances and an agenda of inclusivity, equity and influence amongst communities involved with renewable transitions (Burke and Stephens, 2018).

However, Jason Chilvers and Helen Pallet (2018) observe that the many calls for climate justice and energy democracy, especially by NGOs, labour organisations, and climate justice advocates, are perhaps ill-defined, as they adopt relatively fixed and pre-determined meanings of both 'democracy' and 'public' and attend to aspects of wider energy systems in compartmentalised ways. They point out that 'most sources emphasize bottom-up, civic and community-based empowerment, ownership, and/or control over energy production and consumption' (Chilvers and Pallet, 2018: 2). They stress that decision-making tools 'are based on indicators to judge energy democracy along three dimensions of popular sovereignty, participatory governance and civic ownership' (Chilvers and Pallet, 2018: 3). These participatory processes impose a residual realist standpoint, emphasising definitions and normativities of energy democracy to the exclusion of others and prescribe universal, pre-given evaluative principles (Chilvers and Pallet, 2018).

Alexander Dunlap (2021) is particularly critical of the energy justice and democracy frameworks advanced by the Sustainable Development Goal (SDG) 7 to 'ensure access to affordable, reliable, sustainable and modern energy for all', asserting that the environmental justice and energy justice do not question the trajectory of techno-capitalist development, instead they work to facilitate the inclusion of people. He argues that these participatory processes involving indigenous or non-indigenous local communities are largely uninterested in 'perspectives outside the dominant culture of modernity, industrial development and universal ideas of human rights' (Dunlap, 2021: 5). Likewise, the notion of increasing access for renewable energy technologies (RET) does not critically interrogate the supply web of these technologies and life cycle (Dunlap, 2021; Mejía-Montero et al., 2021): 'while RET may differ from their fossil fuel counterparts, the systems themselves invariably emerge within the same political and cultural economies' (Fortier et al., cited in Mejía-Montero et al., 2021: 1).

Communities get locked into the traditional focus of development planning in renewable energy landscapes (Pasqualetti, 2011). Collective public action and mass movements attempt to challenge power relations and obligations by demanding community ownership, greater participation of the marginalised, energy sovereignty and the creation of green energy/jobs/policies, amongst other things (Chilvers and Pallet, 2018). However, as Ladha and Murphy (2022) stress, the just transition framing is based on the existing polity, i.e. the nation-state and rule of law, while advocating for more equality suggests that elites should give up power. In other words, just transition still legitimates and reifies existing institutions, and while demands are made to change the power relations, existing 'business as usual' systems remain in place. As Crist puts it,

> What remains unstated in the trend of reifications that characterise the Anthropocene discourse (projections of rising human numbers, continued economic development, expanding technological projects and incursions, and deepening the biodiversity crisis) is the abdication of freedom that reifying the trends affirm: the freedom of humanity to choose a different way of inhabiting the Earth is tacitly absent. (Crist, 2016: 22)

In the words of Bayo Akomolafe, 'Demands for social justice get us a seat at the table, but they will never let us leave the house of modernity' (cited in Ladha and Murphy, 2022: 109). This next section explores systemic alternatives offered by the pluriverse, approaches outside the usual neoliberal constructs of the Anthropocene discourse. Arturo Escobar (2020: 53) emphasises the importance of acknowledging epistemologies of the South and the political ontologies of 'contemporary knowledges and struggles orientated towards defence of life and the pluriverse [world views or ontologies for which the world is always multiple]'.

Decoloniality and the Recognition of Indigenous Cosmologies for Climate Change Responses

In this section, I explore the importance of acknowledging and recognising worldviews that are on the defence in social production of life, and how these values and ways of being might assist in responding to the climate crisis differently. While different scholars bring overlapping non-dualistic and divergent perspectives to exploring the 'ontological turn' and de-centring the values of the dominant ontology of Western modernity, my key intention is to make visible the need for alternatives to the 'one-world story' that can displace the 'centrality of this dualist ontology, while broadening the space for non-dualist ontologies' (Escobar, 2020: 49).

Interestingly, Patel and Moore (2020: 1931) point out that 'there isn't a word in English for the process of making life, though such words are found in a range of other languages'. They use the idea of *oikeios* to denote this. In the *oikeios*, humanity is unified with nature in a flow of flows (Moore, 2015b). Patel and Moore (2020: 13) refer to other languages that have concepts of interconnectedness and interrelatedness of life:

> The Anishinaabeg, whose original lands extended widely across north-eastern North America, have 'minobimaatisiiwin', which means 'the good life' but also 'a continuous rebirth' of reciprocal and cyclical relations between humans and other life. Southern African Bantu languages have 'ubuntu', human fulfilment through togetherness, and the Shona language has the further idea of 'ukama', a 'relatedness to the entire cosmos,' including the biophysical world. Similar interpretations exist of the Chinese 'shi-shi wu–ai' and the Maori 'mauri'.

My intention is not to explore indigenous cosmologies in the narrow cultural sense, but to engage with the values that cultivate life-affirming ways as opposed to the version of the human premised on the heteropatriarchal, white supremacist and colonialist social orders prevalent in the world today, perpetuating the production of non-relationality. Borrowing from Imre Szeman and Jennifer Wenzel (2021: 508), who refer to *The Wizard of the Crow* by Kenyan novelist and critic Ngũgĩ wa Thiong'o (2006), which captures anthropocentrism: 'They knew how to take but not to give back to the soil', I strive here to show a politics that recognises the agency of the Other—marginalised, depoliticised, made invisible—and engages with more-than-human worlds (Gibson-Graham, 2014). I argue that a decolonial and relational praxis in response to climate change is needed. Decoloniality, explained by Mignolo (2017:16), is 'delinking from the colonial matrix of power'. It is embodied in relational praxis that disengages from and rejects the colonial structures of power, and in building and being in the world 'otherwise' (Mignolo, 2015). It does not mean the absence of colonialism, which Mignolo (2017) believes may never disappear, but is about a quest to dismantle dualisms and recognise multiple and plural ways of thinking, being, knowing, doing, and sensing. Decoloniality in this sense is delinking from the notions of the Anthropocene discourse.

Relational thinking and being recognise the importance of recalibrating approaches to nature and each other that are restorative, life-affirming, and regenerative, which invites a way of being altogether different from the entrapment of Western modernity (Andreotti, 2006; Andreotti and Menezes de Sousa, 2008; Jimmy and de Oliveira Andreotti, 2021). Relational approaches to researching the interconnectedness of human and non-human worlds draw on indigenous cosmologies, decolonial and relational modes of meaning and world-making. According to Cecilia Åsberg and Rosi Braidotti (2018: 3), relational praxis 'bring[s] things together, new stories and modes of worldly relationality, allowing for their reconfigurations and reconstitutions'. These sometimes-irreverent works debunk economic rationality, science reductionism and an all too 'human' focus, and question 'the universal role, mastery, and nature of human nature' (Åsberg and Braidotti, 2018: 3). I draw inspiration from indigenous cosmovisions, black ecologies, ecofeminism and the more-than-human worlds to bring attention to the importance of decoloniality and relationality. As Davis et al. (2019: 8) argue, 'such humans were responsible for its protection through the enactment of spiritual rituals and other practices to assure its use for collective benefit'.[2]

Latin American social movements and indigenous people offer a mode of thinking that counters capitalist logic and reductionist thinking: *buen vivir*[3]—to

[2] Accordingly, "we must recognise that numerous Black, Brown, and Indigenous people who have, for many years, practised non-binary conceptions of the human and nonhuman relationships" (Davis et al., 2019: 10). These are the humans referred to.

[3] According to Gudynas (2011: 444–445), 'Buen Vivir is not restricted to indigenous postures', and its pluralist endeavour is grounded in ethical perspectives and values that are 'against conventional domination of utilitarian values, particularly expressed in the reductionism of life to economic values and the subsequent commodification of everything'. 'Buen Vivir promotes the dissolution of Society–Nature dualism', argues Gudynas, 'moves away from the prevalence of instrumental and manipulative rationality' and 'as a platform is

live well/fully. *Buen vivir*'s relationship to *Pachamama* (Mother Earth) considers energy as both material and spiritual and is related to indigenous communal organisation and wholeness of land and territory (Mignolo, 2016). In this way of being, humans live in harmony with nature, produce enough for their needs and use local, democratic, participatory approaches. This requires displacing the centrality of humans as the sole subject entitled to political representation, disengaging from economic growth for its own sake and rejecting colonialism and embracing the embodiment of ecocentrism, amongst alternatives to capitalist development. Gudynas (2011) similarly describes the need for transitions that are neither cosmetic nor a return to 'development alternatives', arguing for a radical change and 'alternatives to development'.

Buen vivir has been explored in connection with alternatives to 'development' fixated on growth and human progress (Escobar, 2011; Mignolo, 2016). Living well in harmony with nature draws on the values of South American Indigenous knowledges based on Andean traditions, namely *suma qamaña* in Aymara, Bolivia and *samak kawsay* in Quenchua, Ecuador—both translated into Spanish as *buen vivir*, meaning 'living well' (Mignolo, 2016: 3). It is important to highlight that *buen vivir* has been co-opted by states and corporations alike. The term well-being has been increasingly incorporated in the sustainable development discourse and in the context of 'development' catch-up to improve the lives of those who have been 'left behind'. However, these improvements are synonymous with normativities of Western modernity's way of life.

Gudynas (2011: 444) refers to the work of Simón Yampara, an Aymara sociologist who describes *suma qamaña* as not being restricted to the 'well-being of property ownership and consumption', but as a 'harmonious balance between material and spiritual components' possible in both social and ecological contexts in the production and reproduction of life.

Nabudere's (2011) concept of Afrikology provides insight into the worldview of wholeness and holism held by ancient Africans. He points out that in ancient Egypt, knowledge was understood as 'self-creating in the process of becoming and self-begetting [and] set[ting] the epistemological basis of knowledge creation and its ontological propagation in an organic way' (Nabudere, 2011: 18). Autonomy is central to self-creating living systems, asserts Escobar (2018). In the self-created creator, the process of the emergence of consciousness is interrelated with becoming aware of one's surroundings and environment. In this way what is felt 'in the heart becomes the centre of consciousness and conception of things around'

---

not restricted to a material dimension as it is also a common dimension [and a spiritual dimension]'.

(Nabudere, 2011: 18). Nabudere further explains that reason and logic are the 'basis of sensory information that reports to the heart, and the tongue releases what the heart thinks' (Nabudere, 2011: 18).

Nabudere (2011) also suggests that the African concept of *ubuntu*, 'I am because you are' (Mbiti, 1969), and other African cosmovisions stem from the originality of Ptah, the Egyptian primordial deity of self-creation. Ubuntu 'is premised on very specific understanding of personhood' in which the full development of personhood comes with shared identity and an individual's humanity fostered through a network of relationships (Ogude, 2018: 1). James Ogude (2018: 6) explains that Ubuntu's idea of personhood is not simply about the interdependence between human beings, but also has to do 'with the broader world, physical and spiritual' and in this way personhood is 'a process of being and becoming' in relationship to the material and spiritual worlds.

Escobar (2011: 139) stresses that in 'emphasising the profound relationality of all life, these tendencies [show that] there are indeed relational world views or ontologies for which the world is always multiple—pluriverse'. Escobar suggests that contemporary struggles that reflect the defence and activation of relational communities and world views are not ideological struggles but can be read as 'ontological struggles; they refer to different ways of imagining life, to another mode of existence' (Escobar, 2011: 139). He posits that the emergence of transitional discourses catalyses 'a transition to an altogether different world'—not returning to the past but bringing to light the site of harmony and interconnectedness of human–nature relationality. Furthermore, Escobar (2018) emphasises that these contemporary conjunctions of social and ecological devastation bring to the fore the need for critical inquiries of Western modernity's ontological dominance. Escobar (2018: 4) presents the notion of relationality in terms of ontological design as a means to contribute to a transition from the 'hegemony of modernity's one world ontology to pluriverse of socionatural configurations'. In the context of transition, he proposes 'designs for the pluriverse' as a tool for reimagining and reconstructing local worlds.

I found the concept of 'border thinking' useful for engaging in decolonial thinking about energy democracy from a relational ontological perspective. The essence of border thinking is 'dwelling *in* the borderlands, metaphorically as well as in concrete material terms, to create new cultural and political imaginaries in a position of being-in-between' (Anzaldua in Schulz, 2017: 132). Border thinking assists an understanding of decoloniality as 'delinking from the ways of knowing and the ways of being that trap us into the promises of modernity and the tentacles of coloniality' (Mignolo, 2017: 16). Juxtaposing frontier-making logic with border thinking could assist in creating a decolonial conception of renewable transitions.

However, as Karsten Schulz (2017) proposes, border thinking requires a willingness to engage in mystical narratives that tell us about ourselves and others, about the living and non-living worlds. This does not mean that border thinking is anti-scientific. Rather, it means that border thinking requires us to practice 'epistemic disobedience' by directing attention towards the ideological, geo-political and body-political location of the subject that speaks (Mignolo in Schulz, 2017: 133). Schulz (2017: 133) posits that 'border thinking seems to be a suitable approach for combining different aspects of political ecology and decoloniality with a pluriversal view of more-than-human ontologies'.

From an ecofeminist perspective, the connection between the reproduction of material needs from nature and the production of social life and spiritual relationships is inseparable (Merchant, 1980). Ecofeminist relationality acts to resolve the contradictions between production and reproduction. Eco-feminists share an understanding of how nature relates to energy flows and cycles in which no surplus is produced, and there is an ever-turning embodiment of internal relations and restoration (Salleh, 1997). Focusing on maintaining the 'ecological rhythms in which we are implicated brings new meaning as well as political notions such as "internal security"' (Salleh, 1997: 155). Energy flows and cycles cultivate relationships that reaffirm life, reclaiming planetary cycles of carbon, water, nitrogen and phosphorous; this approach recognises nature as a subject in its own right (Salleh, 1997). Salleh (1997: 56) also draws attention to a 'trans-species' ethic in the understanding of bioenergetics that is useful for our contemporary era, an ethic that encompasses 'an organically living lifespan, birth and death; temporal phasing of organic growth, sexuality, cooperation of meeting of organic needs; stability of social order and integration of social groups'. In terms of relational energy transition, Salleh (1997) explores an understanding of nature's way of relating to energy flows and cycles, making a shift from extractive notions of energy use that underpin *productivist* models to a shared energising force between the human species and the rest of our ecosystems. She writes that these areas are described in almost every culture as 'women's holding labour', thereby giving women a unique vantage point from which to help create an ecocentric ethic (Salleh, 1997: 157).

In terms of relational responses to climate change in energy transitions, I draw on the work of J. K. Gibson-Graham, who sheds some light on considering what more-than-human interdependence and non-capitalist values may look like amongst certain people and communities in Australia. Gibson-Graham (2014: 92) writes that 'there are many more elements of the complex social technical assemblages that are bringing this new energy sector into being'. The language and lexicon of indigenous communities, community economies and sustainable economies 'that articulate the concerns of more-than-human interdependence is an important actant in these world-making preformations as are the collective subjects that are being produced through collaborative and collective practices' (Gibson-Graham, 2014: 92). They provide examples of 'Solar Citizens' and solar co-operatives in Australia, different socio-technical energy assemblages that are non-capitalist and embody more-than-human relations. In the Solar Citizens project, the government programme provided 'cash rebates on the purchasing and installation of solar panels, renewable certificates, and guaranteed energy tariffs, while energy fed the grid' (Gibson-Graham, 2014: 87). These solar panels were placed mainly in households in rural areas and city outskirts through government subsidies (Gibson-Graham, 2014). When the subsidies came to an end, different mechanisms for accessing and financing solar energy emerged. Gibson-Graham (2014: 88) presents an example in which a Californian solar company installed solar panels for free, and arrangements were made 'with a local Credit Union for the system to be paid off over ten years'. In this way the producer and consumer of energy serviced this debt, saving on energy cost as the solar setup produced a surplus of energy that could be sold back into the grid (Gibson-Graham, 2014). The Solar Citizens movement was centred on generating clean energy with collective 'investors taking back control of their energy needs' (Gibson-Graham, 2014: 88). This focused not only on the 'middle class' but all social classes, including those in so-called poorer neighbourhoods. Gibson-Graham argues:

> Solar energy 'prosumers' (producers/consumers) are exercising their calculative agencies that make visible their interdependence with a variety of human and non-human others: the sun and its daily and seasonal rhythms, with the technologies of solar capture and their financing, with the national grid and its demands and price fluctuations, and household energy use. (Gibson-Graham, 2014: 88)

Current frameworks for renewable energy development are 'biased in favour of property-ownership, for-profit business and individuals with economic means to shoulder the up-front cost associated with ownership' (Morris, 2013: 24). Unlike the frameworks that perpetuate 'local neoliberalism' in renewable energy transitions, this solar co-operative project worked based on non-capitalist values.

Solar co-operatives are another interesting post-capitalist initiative launched by labour unions in coal mining areas of Australia (Gibson-Graham, 2014). With the decrease in employment in the manufacturing and heavy industry sectors, a group experimented with setting up a co-operative linked to the renewable energy sector. This group of workers formed 'Australia's first workers' cooperative for "green" manufacturing, called Earthworker Cooperatives' (Gibson-Graham, 2014: 89).

They planned to manufacture solar water units and a full range of 'green' technologies, locating the manufacturing plant in the established coal-mining areas. Gibson-Graham (2014) points out that the business model had an interesting marketing and distribution strategy that combined diverse economies. The solar hot water systems included an 'in-kind payment in negotiation between unions and employers as part of the Enterprise Bargaining Agreements' (Gibson-Graham, 2014: 89). Decisions on surplus distribution were made democratically, including the decision that some of the surplus would be invested back into the co-operative creating more work and a research fund, and a percentage would go towards a social justice project to assist low-income families and older citizens to install solar hot water systems. Gibson-Graham (2014) underscores the fact that this promising initiative by the labour movement had to navigate the complex politics of the workers in the coal sector. They point out that miners are locked into the socio-technical assemblages of exploitative working conditions in a coal mine and appear reluctant to think ahead about the future of the planet. Many would prefer to find jobs at the next mine or factory. In contrast, the Earthworker initiative uses a worker-controlled business model that allows the surplus to be directed towards expanding a sustainable production sector that contributes to the well-being of people and the planet (Gibson-Graham, 2014). The initiative has designed 'socio-technical assemblages that reduce living cost, generate jobs with a liveable wage, care for earth others, and acknowledge our interdependence with future generations' (Gibson-Graham, 2014: 90). He emphasises that 'these kinds of assemblages make community economies more viable' (Gibson-Graham, 2014: 90).

Conclusion

This chapter illuminated the fact that green energy technology responses to the climate crisis are entrenched in the dualist ontology of Western modernity's relations and objectifying epistemology of reductionist and rationalist thought. While post-colonial discourses challenge hegemonic imperial power and knowledge, they largely remain focused on inclusivity in Western modernity's dominant culture. Decolonial and relational scholars centre Cartesian dualisms as problems that underlie an entire structure of institutions and practices through which a Western modernist worldview idea is enacted with the intrinsic separation of things that occludes relationality (Moore, 2015b; Mignolo, 2016; Patel and Moore, 2020; Escobar, 2018). In my exploration of the material of decolonial and relational scholars, recognition was made that numerous black, brown and Indigenous people have practised non-dualist conceptions of human and non-human ways of being, thinking, knowing and doing, based on the profound awareness of interdependence and inter-relatedness of life. Escobar (2018) emphasised that all transitional thinking needs to be attuned to Earth; climate change responses need to consider the multiplicity of entanglements between human and non-human nature based on relational and convivial ways of being, knowing and doing. In recalibrating the relational practice of energy flows and cycles, indigenous cosmologies are essential to foregrounding thinking that will shift the dominant utilitarian understanding of energy needs and use in climate change discourses. The challenge remains of how we might become capable of designing a relational life—and world-making praxis.

In Anthropocene discourse, climate mitigation and adaptation based on relational values and approaches are concealed and undermined in the dominant techno-fix and market-based climate solutions. As Escobar (2011: 139) puts it, 'We need to stop burdening the Earth with the dualisms of the past centuries, and acknowledge the radical interrelatedness, openness, and plurality that inhabit it'. Recalibrating relationships based on life-centric values and principles of reciprocity, restoration and regeneration explored in this chapter, together with the praxis of non-capitalist more-than-human relations, opens possibilities to cultivate a politics of multiplicity and relationality towards more-than-economic value post-extractive energy futures.

References

Andreotti, V. (2006). Soft versus critical global citizenship education. *Policy and Practice: A Development Education Review*, 3, 40–51

Andreotti, V., & Menezes de Sousa, L. M. T. (2008). *Learning to read the world through other eyes*. United Kingdom: Global Education

Åsberg, C., & Braidotti, R. (2018). Feminist post-humanist: An introduction. In C. Åsberg & R. Braidotti (Eds.), *A Feminist Companion to the Posthumanities* (pp. 1–22). Cham: Springer International Publishing. http://dx.doi.org/10.1007/978-3-319-62140-1_1

Baskin, J. (2015). Paradigm dressed as epoch: The ideology of the Anthropocene. *Environmental values*, 24(1), 9–29. http://dx.doi.org/10.3197/096327115X14183182353746

Bond, P. (2012). Emissions trading, new enclosures and eco-social contestation. *Antipode*, 44(3), 684–701. https://doi.org/10.1111/j.1467-8330.211.00890

Bonneuil, C. & Fressoz, J. (2016). *Shock of the Anthropocene: The Earth's history and us*. London: Verso.

Burke, M. J., & Stephens, J. C. (2018). Political power and renewable energy futures: A critical review. *Energy Research & Social Science*, 35, 78–93.https://doi.org/10.1016/j.erss.2017.10.018

Chakrabarty, D. (2009), The climate of history: Four theses. *Critical Inquiry*, 35(2), 197–222. https://doi.org/10.1086/596640

Chilvers, J., & Pallett, H. (2018). Energy democracies and publics in the making: A relational agenda for research and practice. *Frontier in Communication*, 3(14), 1–16 http://dx.doi.org/10.3389/fcomm.2018.00014

Climate Justice Alliance (n.d.). *About.* Retrieved December 9, 2024, from https://climatejusticealliance.org/about/

Crist, E. (2016). On the poverty of our nomenclature. In J. W. Moore (Ed.), *Anthropocene or Capitalocene? Nature, history and the crisis of capitalism.* Oakland CA: PM Press

Cronon, W. (1983). *Changes in the land: Indians, colonists, and the ecology of New England*(1st ed). York: Hill and Wang

Crutzen, P., & Schwägerl, C. (2011). Living in the anthropocene: Towards a new global ethos. *Yale Environment 360.* Retrieved May 1, 2025, from https://e360.yale.edu/features/living_in_the_anthropocene_toward_a_new_global_ethos

Dalsgaard, S. (2013). The commensurability of carbon: Making value and money of climate change. *Journal of Ethnographic Theory*, 3(1), 80–98 https://doi.org/10.14318/hau3.1.006

Davis, J., Moulton, A. A., Van Sant, L., & Williams, B. (2019). Anthropocene, Capitalocene, Plantationocene? A manifesto for ecological justice in an age of global crises. *Geography Compass*, 13(5). https://doi.org/10.1111/gec3.12438

De la Cadena, M. (2019). Uncomming nature stories from the anthropo-not-seen. In Harvey, P; Krohn-Hansen, C and Nustad, K, (eds), *Anthropos and the Material.* Duke University Press.

Dunlap, A. (2021). More wind energy colonialism(s) in Oaxaca? Reasonable findings, unacceptable development. *Energy Research & Social Science*, 82. https://doi.org/10.1016/j.erss.2021.102304

Ehrlich, P. R. (1971). *The population bomb. Rev.* (Ed.). New York: In Text

Escobar, A. (1996). Construction Nature. Elements for a Post-Structuralist Political Ecology. *Futures* 28(4),325–353.

Escobar, A. (2011). Sustainability: Design for the pluriverse. *Development*, 54, 137–140

Escobar, A. (2018). *Designs for the pluriverse: Radical interdependence, autonomy, and the making of worlds.* Durham: Duke University Press. https://doi.org/10.1215/9780822371816

Escobar, A. (2020). *'Thinking–feeling with the Earth: Territorial struggles and the ontological dimension of the epistemologies of the south', in knowledges born in the struggle.* (1st ed [Online]). Routledge. pp. 41–57.

Fay, J. (2013). *The role of market-based incentives in promoting low carbon development in developing countries.* PhD Thesis. University of Cape Town.

Finon, D. (2013). The transition of the electricity system towards decarbonization: The need for change in the market regime. *Climate Policy*, 13(1), 130–145.

Fioramonti, L. (2014). *How Numbers Rule the World: The Use and Abuse of Statistics in Global Politics.* Zed Books. https://doi.org/10.5040/9781350220645

French, K. B., Sanchez, A., & Ullom, E. (2020). Composting settler colonial distortions: Cultivating critical land-based family history. *Genealogy*, 4(3). https://doi.org/10.3390/genealogy4030084

Gabor, D. (2021). The Wall Street Consensus. *Development and change*, 52(3), 429–459. https://doi.org/10.1111/dech.12645

Galgóczi, B. (2020). Just transition on the ground: Challenges and opportunities for social dialogue. *European Journal of Industrial Relations*, 26(4), 367–382. https://doi.org/10.1177/0959680120951704

Gibson-Graham, J. K. (2014). Being the revolution, or, how to live in a 'more-than-capitalist' world threatened with extinction. *Rethinking Marxism*, 26(1), 76–94

Goodland, R., & Daly, H. (1996). Environmental sustainability: Universal and non-negotiable. *Ecological Applications*, 6(4), 1002–1017

Green Climate Fund (n.d.). *Private sector financing*. Retrieved December 9, 2024, from https://www.greenclimate.fund/sectors/private

Gudynas, E. (2011). Buen Vivir: Today's tomorrow. *Development*, 54(4), 441–447

Haraway, D., 2016. Staying with the trouble: Making Kin in the Chthulucene. Durham, NC: Duke University Press. https://doi.org/10.1215/9780822373780

Howe, C. (2014). Anthropocenic ecoauthority: The Winds of Oaxaca. *Anthropological Quarterly*, 87(2), 381–404. https://doi.org/10.1353/anq.2014.0029

Howe, C. (2019). *Ecologics*. Durham, NC: Duke University Press.

International Energy Agency. (2017). Energy access outlook report 2017. *World Outlook Special Report*. Retrieved June 20, 2018, from https://www.iea.org/reports/energy-access-outlook-2017

IPCC (n.d.). *About the IPCC*. Retrieved December 9, 2024, from https://www.ipcc.ch/about/

IPCC. 2007, Climate Change 2007: *Mitigation of Climate Change. Contribution of Working Group III to the Fourth Assessment Report of the Intergovernmental Panel on Climate Change*. [B. Metz, O.R. Davidson, P.R. Bosch, R. Dave & L.A. Meyer (Eds.)]. Cambridge University Press, Cambridge. Retrieved May 1, 2025, from https://www.ipcc.ch/report/ar4/wg3/

IPCC (2022). *Climate change 2022: Mitigation of climate change. Contribution of working group III to the sixth assessment report of the intergovernmental panel on climate change*. [P.R. Shukla, J. Skea, R. Slade, A. Al Khourdajie, R. van Diemen, D. McCollum, M. Pathak, S. Some, P. Vyas, R. Fradera, M. Belkacemi, A. Hasija, G. Lisboa, S. Luz & J. Malley, (Eds.)]. Cambridge & New York, NY: Cambridge University Press

Jimmy, E., & de Oliveira Andreotti, V. (2021). Conspicuous Consumption: Economies of virtue and the commodification of Indigeneity. *Public*, 32(64), 121–133

Khare, R. (2016). The principle of 'Common but Differentiated Responsibilities' and the challenges posed by it in the context of international climate governance. *International Journal of Law and Legal Jurisprudence Studies*, 3(2), 98–113.

Ladha, N., & Murphy, L. (2022). *Post capitalism philanthropy: Healing wealth in a time of collapse*. Daraja Press and Transition Resource Circle

Mbiti, J. S. (1969). *African religions and philosophy*. Nairobi: East African Educational Publishers Ltd.

Mejía-Montero, A., Lane, M., van Der Horst, D. & Jenkins, K. E. H. (2021). Grounding the energy justice lifecycle framework: An exploration of utility-scale wind power in Oaxaca, Mexico. *Energy Research & Social Science*, 75, 102017. https://doi.org/10.1016/j.erss.2021.102017

Merchant, C. (1980). *The death of nature: Women, ecology, and the scientific revolution.* New York: Harper & Row

Mies, M. (2014). The myth of catching up development. In M. Mies, & V. Shiva (Eds.), *Ecofeminism* (pp. 55–69). London: Zed Books

Mignolo, W. (2011). *The darker side of Western modernity: Global futures, decolonial options*. Durham, North Carolina: Duke University Press

Mignolo, W. (2015). *The role of BRICS countries in the becoming world order: "Humanity," colonial/imperial differences, and the racial distribution of capital and knowledge*. Conference Presentation

Mignolo, W. (2016). *Sustainable development or sustainable economies? Ideas towards living in harmony and plentitude*, DOC Research Institute, 2016: https://doc-research.org/

Mignolo, W. (2017). Interview with Walter D. Mignolo, Interview was conducted by Sebastian Weier, In Woons, M. and Weier, S (eds), *Critical Epistemologies of Global Politics*. E-International Relations Publishing, Bristol, England.

Mitchell, T. (2009). Carbon democracy. *Economy and society*, 38(3), 399–432

Moore, J.W. (2002). The Crisis of feudalism: An environmental history. *Organization & Environment*, 15 (3): 301–322.

Moore, J. W. (2014). The end of cheap nature or. How I learned to stop worrying about 'the' environment and love the crisis of capitalism. In C. Suter & C. Chase-Dunn (Eds.), *Structures of the World Political Economy and the Future of Global Conflict and Cooperation*. LIT Verlag

Moore, J. W. (2015a). Putting nature to work. In C. Wee, J. Schönenbach, & O. Arndt (Eds.), *Supramarkt: A micro-toolkit for disobedient consumers, or how to frack the fatal forces of the Capitalocene* (pp. 69–117). Gothenburg: Irene Books

Moore, J. W. (2015b). *Capitalism in the web of life: Ecology and the accumulation of capital*. London: Verso Books

Moore, J. W. (2016). The rise of cheap nature. In J. W. Moore (Ed.), *Anthropocene or Capitalocene? Nature, History, and the Crisis of Capitalism* (pp. 78–116). Oakland: PM Press

Moore, J. (2017). The Capitalocene, Part I: On the nature and origins of our ecological crisis. *The Journal of Peasant Studies*. https://doi.org/10.1080/03066150.2016.1235036

Moore, J.W. (2021). *"Climate, class & the great frontier: From primitive accumulation to the. Great implosion,"* Unpublished Paper, World–Ecology Research Group. Retrieved 10 July 2022 from https://jasonwmoore.com/wp-content/uploads/2021/09/Moore-Class-Climate-and-the-Great-Frontier-circulating-September-2021.pdf

Morris, J. (2013). The evolving localism (and neoliberalism) of urban renewable energy projects. *Culture, Agriculture, Food and Environment*, 35(1), 16–29

Nabudere, D. (2011). *Afrikology, philosophy and wholeness: An epistemology*. The Africa Institute of South Africa

Ogude, J. (2018). Introduction: Ubuntu and personhood. In J. Ogude (Ed.), *Ubuntu and personhood*. Trenton, New Jersey: Africa World Press

Pasqualetti, M. J. (2011). Social barriers to renewable energy landscapes. Geographical Review. 101(2). 201–223

Patel, R., & Moore, J. W. (2020). *A history of the world in seven cheap things: A guide to capitalism, nature and the future of the planet*. London, UK: Verso. https://doi.org/10.1525/9780520966376

Pollitt, M G. (2012). The role of policy in energy transitions: Lessons from the energy liberalisation era. *Energy Policy*, 50, 128–137.

Quijano, A. (1991/2007). Coloniality and modernity/rationality. *Cultural Studies*, 21(2–3), 168–78

Salleh, A. (1997). *Ecofeminism as politics: Nature, marx and the postmodern*. London: Zed Books

Salleh, A. (2010). Climate strategy: Making the choice between ecological modernisation or living well. *Journal of Australian political economy*, 66(66), 118–143

Schulz, K. A. (2017). Decolonizing political ecology: Ontology, technology and 'critical' enchantment. *Journal of Political Ecology*, 24(1), 125–143. https://doi.org/10.2458/v24i1.20789

Steffen, W., Cruzen, P., & McNeill, J., (2007). The Anthropocene: Are humans now overwhelming the great forces of nature?. *Ambio*, 36(8), 614–621

Stengers, I. (2015). Accepting the reality of Gaia – A fundamental shift? In C. Hamilton, F. Gemenne, & C. Bonneuil (Eds.), *The Anthropocene and the Global Environmental Crisis: Re-thinking modernity in a new epoch* (pp. 134–144). Routledge

Stevis, D. & Felli, R. (2015). Global labour unions and just transition to a green economy. *International Environmental Agreements: Politics, Law and Economics*, 15(1), 29–43

Sweeney, S. & Treat, J. (2018). Trade unions and a just transition: A search for a transformative politics. *Trade Unions for Energy Democracy*. Retrieved May 1, 2025, from https://www.tuedglobal.org/working-papers/trade-unions-and-just-transition-the-search-for-a-transformative-politics

Swilling, M., & Annecke, E. (2012). *Just transitions: Explorations of sustainability in an unfair world*. South Africa: UCT Press

Szeman, I., & Wenzel, J. (2021). What do we talk about when we talk about extractivism? *Textual Practice*, *35*(3), 505–523. https://doi.org/10.1080/0950236X.2021.1889829

Tsing, A. (2012, November). Unruly edges: Mushroom as companion species. *Environmental Humanities*, 1, 141–154. https://doi.org/10.1215/22011919-3610012

UNFCCC. (n.d.). *International Emissions Trading*. Retrieved May 1, 2025, from https://unfccc.int/international-emissions-trading

wa Thiong'o, N. (2006). *Wizard of the crow*. New York: Anchor Books

World Commission on Environment and Development (WECD)(1987). Towards Sustainable Development. In Our Common Future. London: Oxford University Press.

Climate Change Knowledge and Epistemic Injustice in a Postcolonial Context

5

Aïda C. Terblanché-Greeff

Abstract

Climate change knowledge is mainly created within and disseminated from a Western scientific paradigm, using internationally dominant language(s) (primarily English) and subject jargon. With the predominantly unidirectional flow of such knowledge from the Global North, limited attention is paid to context-specific climate change knowledge created in the Global South and to whether and how such knowledge is to be disseminated. This chapter is the first to argue that this lack of recognition with respect to thought about climate change in some of the developing countries in the Global South consists of epistemic injustice—roughly, unjust treatment relating to issues pertaining to knowledge production and dissemination. In particular, it argues that two types of epistemic injustice regarding climate change knowledge, viz., hermeneutical and contributory injustice, are prevalent in the postcolonial context of a developing country like South Africa. To correct this epistemic injustice, the chapter proposes concrete ways of decolonising climate change knowledge that would involve substantial shifts in the way climate change mitigation and adaptation strategies and policies are created and communicated.

A. C. Terblanché-Greeff (✉)
North-West University, Potchefstroom, South Africa
e-mail: Aida.Botha@nwu.ac.za

A. C. Terblanché-Greeff et al. (eds.), *Climate Change and the Postcolonial*, Kulturelle Figurationen: Artefakte, Praktiken, Fiktionen,
https://doi.org/10.1007/978-3-658-49328-8_5

Keywords

Climate change epistemic injustice · Climate change knowledge · Contributory injustice · Hermeneutical injustice · Postcolonial

Introduction

Earth's climate is changing at a record rate to such a degree that humankind and multitudinous species will eventually face extinction should the rate of human-induced climate change continue without drastic mitigation and adaptation measures being put into place. However, not everyone will be equally impacted by climate change in the immediate future. For example, African countries are some of the lowest greenhouse gas emitters compared to more developed countries. Yet, they are exponentially affected by climate change across various developmental sectors (Trisos et al., 2022). As Christopher H. Trisos and his colleagues (2022: 1290) point out, '[m]ost African countries will enter unprecedented high temperature climates earlier in this century than generally wealthier, higher latitude countries, emphasising the urgency of adaptation measures in Africa'. There will be, so to speak, degrees of climate change losers based on different states of vulnerability and capacity when faced with hazards.[1] With this in mind, it becomes crucial that climate change knowledge be created and disseminated on a glocal scale—that is, both globally and locally (in a nuanced manner)—and context-specific climate change mitigation and adaptation strategies must be developed.

The most prominent instance of climate change knowledge creation and dissemination is through the assessment reports published by the Intergovernmental Panel on Climate Change (IPCC).[2] Established in the 1980s, the IPCC is an international scientific body that assesses existing empirical research related to climate change to provide governments and policymakers with information to be used during various activities, such as international negotiations. Furthermore, the assessment reports are considered a reliable source of scientific climate change knowledge. They are most often used to assist governments in making informed

[1] Here, the vulnerabilities and capacities will depend on geographical location, material wealth, and global agency, to name a few contributing factors.

[2] These different reports cover various topics, such as 'The physical science basis', 'Impacts, adaptation and vulnerability', and 'Mitigation of climate change'.

decisions regarding climate change and its impacts when developing climate change mitigation and adaptation strategies and policies.

Climate change knowledge can be broadly described as information on the Earth's changing climatic patterns and the causes and effects of these changes. It also includes information on vulnerability, impacts, mitigation, and adaptation related to climate change. Moreover, climate change knowledge stems from multiple epistemologies—that is, knowledge pertaining to climate change should not be considered as limited to modern scientific knowledge but should also encompass other cultural and social epistemologies like indigenous and local knowledges. This is supported by the IPCC (2023: 7) when it states that '[t]his report [viz., the 6th Assessment Report] recognises the value of diverse forms of knowledge such as scientific, as well as indigenous knowledge and local knowledge in understanding and evaluating climate adaptation processes and actions to reduce risks from human-induced climate change'. More specifically, the United Nations Educational, Scientific and Cultural Organization (UNESCO, 2021) states that '[l]ocal and indigenous knowledge refers to understandings, skills and philosophies developed by societies with long histories of interaction with their natural surroundings.' For rural and indigenous peoples, local knowledge informs decision-making about fundamental aspects of day-to-day life. This knowledge is integral to a cultural complex that also encompasses language, systems of classification, resource use practices, social interactions, ritual and spirituality. These unique ways of knowing are important facets of the world's cultural diversity and provide a foundation for locally appropriate sustainable development.

From the preceding, it becomes obvious that distinct climate change knowledges—based on various epistemologies—are cardinal in humankind's ability to mitigate and adapt to the effects of climatic changes faced by all, specifically the most vulnerable regions, like Africa, South America, etc. Governments in the Global South use climate change knowledge to develop their climate change mitigation and adaptation strategies and policies. Ideally, these strategies and policies should be context-specific and nuanced, which is not always the case. Instead, such strategies and policies are frequently conclusively based on knowledge disseminated from developed countries in the Global North.

However, there is often an overreliance by governments and policymakers on globally recognised scientific reports like that of the IPCC, which results in climate change knowledge being primarily created and disseminated from a Western perspective and by using dominant international languages (e.g., English) and subject jargon. In fact, it is repeatedly the case that climate change mitigation and adaptation strategies and policies of developing countries in the Global South are created using the Western technocratic 'mould' which stems from the modern scientific episteme and its successful globalisation.[3] This is prevalent in various countries—for example, South Africa—where the governments acknowledge that indigenous and local knowledge should be included in the creation and dissemination of climate change knowledge. However, the South African government does not always indicate precisely *how* this is to be done (see National Planning Commission, 2012). Instead, matters regarding climate change knowledge—such as the development of climate change mitigation and adaptation strategies and policies—are often reserved for specific sectors of society, viz., scholars, politicians, industry stakeholders, non-governmental organisations, etc. Some groups (e.g., rural populations with indigenous and local knowledge) are frequently excluded from climate change discursive practices as epistemic subjects, indicating that some governments disregard and disadvantage some groups as *knowers* of climate change knowledge (albeit to various degrees) (Byskov and Hyams, 2022).[4] This can lead to climate change epistemic injustice.

It is useful to unpack the concept of climate change epistemic injustice. Here, epistemic injustice can be explained as the 'unfair treatment that relates to issues of knowledge, understanding, and participation in communicative practices' (Kidd et al., 2017: 1). With this in mind, climate change epistemic injustice may be described as a wrong done to someone as a knower or transmitter of climate change knowledge. An example of climate change epistemic injustice is when someone—from a marginalised group—is oppressed and excluded from (co-) producing or disseminating climate change knowledge.[5]

---

[3] This is exemplified by South Africa's Climate Change Bill B9-2022, which states that, 'The interpretation and application of this Act must be guided by [...] the need for climate change mitigation and adaptation responses to be informed by evolving climate change scientific knowledge and decisions which should be based on the best available science, evidence and information' (Republic of South Africa, 2022: 7).

[4] This chapter follows the lead of Morten F. Byskov and Keith Hyams (2022) in its approach to the subjects of epistemic injustice. Accordingly, '[a] relevant question here is whether an epistemic injustice is related to the treatment of Indigenous peoples as individual knowers or whether it concerns Indigenous knowledges as bodies of knowledge [...] we here hold that an epistemic injustice is committed against [a] person (as an individual knower) or [a] community (as a collective knower). When we refer to Indigenous knowledges as subject to epistemic injustice, what we mean is that such knowledge is in the abstract often excluded from decision-making processes, but that such exclusion in practice takes the form of epistemically excluding [...] communities and individuals from the processes.' (Byskov and Hyams, 2022: 618).

[5] Due to limited space in the current chapter, an in-depth discussion regarding knowledge production and the long-established strand of the history and anthropology of sciences and

This chapter is divided into several sections. Before it is argued that climate change epistemic injustice is prevalent in a postcolonial context (specifically, South Africa), it will be useful to explain in more detail what the concept of epistemic injustice means. Epistemic injustice is elaborated on through two types: hermeneutical and contributory. After that, some of the colonial remnants pertaining to climate change knowledge are highlighted before turning to the postcolonial South African context. Epistemic *othering* is briefly discussed to elucidate the prevalence of climate change epistemic injustice in this postcolonial setting. Finally, a path forward is proposed: addressing the epistemic injustice prevalent in climate change knowledge through decolonisation. This requires major changes in how climate change mitigation and adaptation strategies and policies are created in a developing country like South Africa.

Epistemic Injustice

Humans mostly have the capacity to engage in social practices as both knowers and seekers of knowledge.[6] According to Jules Holroyd and Kathy Puddifoot (2020), humans engage as epistemic subjects in three ways. Firstly, there is an interchange between humans where knowledge is received and shared with others. Secondly, humans try to acquire more knowledge and use it to make sense of our *being-in-the-world* and *being-with-others*. Thirdly, humans attempt to contribute and be recognised as knowledge creators, thus adding to our shared understanding of the world and our being in it.

However, not all knowers have the option to create or exchange knowledge equally.[7] That is, some knowers are considered to be superior to others. Here, our ways of seeking and sharing knowledge are unjust as some epistemic subjects are considered lesser (often due to a lack of recognition). Epistemic subjugation becomes prevalent between various actants.

Philosopher Miranda Fricker (2007) brings together epistemology and ethics by presenting *epistemic injustice* as a concept. According to her, epistemic injustice can be defined as a particular form of wrongdoing 'done to someone specifically in their capacity as a knower' (Fricker, 2007: 1). She identifies

knowledge cannot be presented. The reader is referred to the work of Knorr-Cetina (1999), Jasanoff (2004), and Rheinberger (2010).

[6] Excluded here are obvious instances where people cannot entirely engage in social practices and where their capacity as knowers is limited, such as people in vegetative states.

[7] In this chapter, the concepts of *knower* and *epistemic subject* are used interchangeably.

various elements that can contribute to epistemic injustice: power relations, oppression, discrimination and distribution. More specifically, 'the root cause of epistemic injustice is structures of unequal power' (Fricker, 2007: 8); it is (often) rooted in an imbalance between the political and the social. She continues by stating that '[t]he world is "structured" by the powerful to the detriment of the powerless, there is another species of oppression at work, one that has not been registered in mainstream epistemology: epistemic oppression' (Fricker, 1999: 191). Kristie Dotson (2012: 36) describes epistemic oppression:

> [It] is primarily characterized by detrimental exclusions from epistemic affairs. Whether concerning hermeneutical resources or discourse on an important matter of social policy, epistemic oppression concerns routine and harmful exclusions from some domain of knowledge production. All the forms of epistemic injustice introduced here involve some form of pervasive, harmful epistemic exclusion. As such, they are all species of epistemic oppression.

Epistemic oppression and the resultant epistemic injustice can take on various guises. In this chapter, attention will briefly be turned to *hermeneutical injustice* (as described by Fricker, 2007) and *contributory injustice* (presented by Dotson, 2012). By investigating these types of epistemic injustice, we can better elaborate on how various concepts pertaining to the postcolonial context and the prevalent climate change epistemic injustice are related to indigenous and local knowledge.

Hermeneutical Injustice

Hermeneutical injustice is 'the injustice of having some significant area of one's social experience obscured from collective understanding owing to persistent and wide-ranging hermeneutical marginalisation' (Fricker, 2007: 154). Moreover, hermeneutical injustice occurs when 'a subject who is already hermeneutical[ly] marginalised (that is, they belong to a group which does not have access to equal participation in the generation of social meaning) is thereby put at an unfair disadvantage when it comes to making sense of significant areas of their social experience' (Fricker, 2013: 1319). There is a structural-identity prejudice as a result of socio-epistemic structures that 'create and sustain situated hermeneutical inequality' (Dotson, 2012: 30).

This becomes evident when some groups are epistemically dominant in that they have the ability to assign epistemic resources to their own social experiences. An example would be the jargon used in climate change discourses based on the dominant Western paradigm that is marked by scientism. In contrast, marginalised others—such as people with indigenous and local knowledge—remain excluded from such meaning-making based on structural inequalities. This subordination has a moral-political aspect, as epistemic inclusion would have value for marginalised groups (Fricker, 2006). The power imbalance becomes evident. Knowers from the epistemic dominant group(s) have more hermeneutical power and are hermeneutically advantaged compared to those from the marginalised group.

Additionally, hermeneutical injustice is marked by a lack of epistemic agency on the part of the marginalised knower due to epistemic exclusion from the dominant group. An example of a hermeneutically marginalised group is people of colour during the apartheid era in South Africa, when this group did not have epistemic agency or power and was hermeneutically disadvantaged.[8] In other words, members of particular social identity groups—that are marginalised—are excluded from constructing shared understandings of social experiences (Fricker, 2007). From this, it is important to note that historical patterns of being hermeneutically marginalised are a prerequisite for suffering epistemic injustice as a knower.

More specifically, such an injustice occurs when the disadvantaged epistemic subject experiences a hermeneutical lacuna when trying to express themselves as situated knowers due to a lack of conceptual resources. Accordingly, '[w]hen gaps in collective hermeneutical resources create significant and potentially harmful obstacles to diagnosing and understanding one's own experiences, one is encountering hermeneutical injustice' (Dotson, 2012: 30). Language becomes important as the lack of the correct linguistic resources (for example, the understanding of jargon-specific linguistics or lack of competence in the lingua franca, for instance English) can hinder the knowers from effectively articulating and expressing their lived experiences. From this it follows that the epistemic subjects' knowledge and experience may be deemed incoherent and misunderstood by the knowers from the dominant epistemic group.

It would, however, be incorrect to reduce all available hermeneutical resources to those of the epistemic dominant group. That is, 'we do not *all* depend on the same hermeneutical resources. Such an assumption fails to take into account *alternative* epistemologies, counter-mythologies, and hidden transcripts that exist in hermeneutically marginalized communities among themselves.' (Dotson, 2012: 31, emphasis added). This then brings us to another form of epistemic injustice: one that is contributory in nature.

[8] Here, the term *people of colour* refers to all racial groups that are not white and, as such, include African, multiracial, Indian, Asian, etc.

Contributory Injustice

In contributory injustice, there is wilful hermeneutical ignorance on the part of the epistemic dominant group to recognise and incorporate other, albeit non-dominant, epistemic resources. Dotson (2012: 32) elaborates on contributory injustice as 'the circumstance where an epistemic [advantaged] agent's willful hermeneutical ignorance in maintaining and utilizing structurally prejudiced hermeneutical resources thwarts a [disadvantaged] knower's ability to contribute to shared epistemic resources within a given epistemic community by compromising her epistemic agency.'

One way of doing such an injustice is being deliberately ignorant of indigenous and local knowledge pertaining to climate change. An example of indigenous and local knowledge in South Africa would be local subsistence farmers in the Free State province using the appearance of red ants in November and December to predict good rains during the summer season, prompting them to prepare for a late sowing period (Zuma-Netshiukhwi et al., 2013). However, this instance of indigenous and local climate change knowledge is not necessarily recognised by the dominant paradigm as it is not explained scientifically. This injustice occurs when the advantaged epistemic group intentionally refuses 'to acknowledge and acquire the necessary tools for knowing whole parts of the world' (Pohlhaus, 2011: 15). This wilful hermeneutical ignorance, paired with a lack of recognition of marginalised group's indigenous and local knowledge systems, can have dire effects on understanding climatic changes (and possible adaptation strategies) applicable to a specific area.

From the preceding, it becomes clear that the local farmers can make sense of and articulate their being-in-the-world. There is not a linguistic lacuna as prevalent in hermeneutical epistemic injustice. Instead, the meaning allocated to social experience is not given appropriate uptake because of the 'biased hermeneutical resources' used by the advantaged epistemic group (Dotson, 2012: 32). Epistemic agency of the marginalised subject is compromised (Dotson, 2012). Accordingly, the Western paradigm (with its focus on scientism) excludes 'certain knowledge subjects or systems simply because they do not meet its criteria of what counts as knowledge, its processes of legitimation' (Posholi, 2020: 296).

Based on the epistemically oppressive nature of practices whereby people of colour were marginalised, subjugated and not recognised as epistemic subjects, it can be argued that epistemic injustice can be discussed within the context of the colonial and postcolonial. Supportively, Rajeev Bhargava (2013: 414) states that epistemic injustice '[...] occurs when the concepts and categories by which a people understand themselves and their world [i.e. being-in-the-world and being-with-others] are replaced or adversely affected by the concepts and categories of the colonisers'.

Colonial Remnants and Climate Change Knowledge

Earlier in this chapter, it was mentioned that historical patterns of being hermeneutically marginalised are a prerequisite for suffering epistemic injustice as a knower. Examples of such historical patterns of harm are patriarchy and colonisation.[9] Here, '[a]n adversely affected site of this sort of historical harm is sub-Saharan Africa through colonialism' (Tobi, 2022: 338). In various postcolonial contexts (such as post-apartheid South Africa), unequal socio-epistemic structures are remnants of colonialism that still linger, albeit subtly and less obviously. This (subtle) coloniality is the 'long-standing patterns of power that emerged as a result of colonialism, but that define [...] knowledge production well beyond the strict limits of colonial administrations' (Maldonado-Torres, 2007: 243).

This form of coloniality spills over into climate change knowledge production in the postcolonial context. In their article titled *Geographical imbalances and divides in the scientific production of climate change knowledge*, Maya Pasgaard and colleagues conducted a bibliometric study (on > 15,000 publications) of the global supply of climate change knowledge (Pasgaard et al., 2015). The results indicate that the 'production of climate change knowledge is biased away from developing, more vulnerable regions of the world with warmer climates and low climate footprints [...]'. (Pasgaard et al., 2015: 280). More explicitly, climate change knowledge production predominantly 'takes place in richer, institutionally well-developed countries with cooler climates and high climate footprints' (Pasgaard et al., 2015: 286). Such climate change knowledge is included in scientific reports, like the IPCC's, which rightly identifies a 'lack of geographical balance in the data and literature on observed changes in natural and managed systems, with a marked scarcity from developing countries' (Rosenzweig et al., 2007: 117).

[9] Although Fricker (2007, 2013, 2017) never mentions the remnants of colonialism as an example of systemic issues that lead to epistemic injustice, it can be argued that such issues are nonaccidental, thus they can be included in the category of epistemic injustice (Cummings et al., 2023).

Consequently, climate change knowledge disseminated by global climate change governance bodies (e.g., United Nations Framework Convention on Climate Change [UNFCCC], etc.) may be less 'global' than anticipated. Sylvia Karlsson, Tanja Srebotnjak, and Patricia Gonzales (2007: 680) elaborate that such knowledge is 'intended to influence policy at global but also at lower governance levels. This means that global negotiations and policy are informed by what may be inappropriate Northern biased "globalized" knowledge'. The subtle coloniality of global climate change knowledge becomes apparent. Furthermore, this imbalance in knowledge production may influence the utilisation and integration of indigenous and local knowledge to provide context-specific advice (Karlsson et al., 2007).

This lack of integration of indigenous and local knowledge into scientific climate change knowledge to inform local climate change mitigation and adaptation strategies and policies can be seen as a form of wilful hermeneutical ignorance. That is ignorance on the part of the national governments of developing countries that use biased climate change information in their local climate change governance. In such cases, an epistemic injustice occurs. Scientific climate change knowledge, as disseminated by the Global North, has become the dominant epistemic resource pool often used by national governments. The hermeneutical marginalised groups with indigenous and local knowledge are excluded from contributing to the resource pool, regardless of their potential contributions as epistemic subjects. Epistemic oppression is prevalent as this practice hinders specific knowers' contributions to knowledge production (Dotson, 2014).

Given the remnants of coloniality prevalent in postcolonial societies, it will be useful to examine the South African context as it pertains to climate change epistemic injustice.

The Postcolonial South African Context

According to Fricker (2007), epistemic injustice cannot be divorced from systemic structures of oppression, domination, inequality, and socio-economic distribution. Such structures are prevalent in the South African postcolonial context and contribute to existing epistemic injustice. One example is the fact that there is an enormous socio-economic distribution divide between various groups in South Africa, with the country having the highest income inequality globally, according to the Gini Index (see Sulla et al., 2022). This is exemplified by the 32.1% unemployment rate in South Africa as of 2023 (Statistics South Africa [STATSSA], 2024).

Socio-economic factors are cardinal in establishing epistemic advantaged and disadvantaged groups. For example, because of socio-economic factors, not all groups have access to educational resources to become proficient in English to such a degree that they will no longer be hermeneutically marginalised when it comes to creating and sharing climate change knowledge. Furthermore, due to socio-economic factors, the lack of epistemic credibility given to rural communities (with indigenous and local knowledge) leads to systematic epistemic exclusion, hermeneutical marginalisation, and a lack of opportunities to contribute to public discourse on climate change knowledge (see Byskov and Hyams, 2022).

Another way to hermeneutically marginalise a group is through structural-identity prejudice that 'exists within collective hermeneutical resources themselves' (Dotson, 2012: 29). More specifically, the sense-making resources of the dominant group with an epistemic advantage may become 'discriminatory due to an asymmetrical ability of some groups to affect the ways in which a given society makes sense of the world' (Dotson, 2012: 29). Again, an example of discrimination caused by a systemic injustice is the use of language—specifically, a lingua franca like English (which might be considered as the most prominent language used in the Western paradigm). This is because language can and has been used to influence power dynamics where there is a hierarchy between the hermeneutical dominant group and the marginalised group(s) (see wa Thiong'o, 1987). Furthermore, Gaile Pohlhaus (2011: 718) explains the importance of language when she states that '[k]nowing requires [epistemic resources], such as language to formulate propositions, concepts to make sense of experience, procedures to approach the world and standards to judge particular accounts of experience'.

The epistemically marginalised groups are 'often required to use a language and a set of assumptions when communicating with [the epistemically dominant group(s)] that do not adequately account for the full range of their experiences' (Dotson, 2014: 127). Marginalised knowers might also be less confident when expressing themselves using the language of the hermeneutical dominant group (Byskov and Hyams, 2022). Even if they can express themselves adequately, 'successful integration of [i]ndigenous knowledges in climate adaptation policy and planning requires recognition of how the current climate language is skewed towards Western scientific and technocratic approaches and how this may limit the input that vulnerable [communities with indigenous and local knowledge] are able to give' (Byskov and Hyams, 2022: 629).

The following example clarifies the socio-economic and linguistic injustices in postcolonial South Africa and their relation to climate change knowledge. One group that has been hermeneutically marginalised and subsequently dehumanised is Black Africans living in South Africa's rural areas who often have low socio-economic status.[10] That is, people who live close to nature, who are dependent on subsistence farming and who have indigenous and local knowledge of their natural environment. They have predominantly been excluded as knowers regarding climate change knowledge when the South African national government develops climate change mitigation and adaptation strategies and policies.[11] This may be due to unequal socio-epistemic structures exacerbated by race, economic class, and educational resources (e.g., access to education leading to language competencies—specifically, competence in English). This exclusion becomes pertinent when we consider the following. Firstly, in rural areas, only 16.1% of South Africans speak English as their primary language at home (Hofmeyr, 2024). Secondly, most, if not all, official South African governmental documents regarding climate change are published in English only—the 'global language'. In a multi-cultural country like South Africa, with 12 official languages, this exclusion creates unwanted power relations between those with hermeneutic power and those who are hermeneutically marginalised regarding climate change knowledge. This leads to epistemic alienation and oppression, as well as disproportionate hermeneutic power. That said, language creates an obstacle to overcome when considering who can contribute to climate change knowledge to be used in climate change mitigation and adaptation strategies and policies in a country like South Africa.

From the preceding, it becomes evident that not everyone is considered an equal contributor regarding climate change knowledge in South Africa. Since much of the climate change knowledge, under the guise of scientism, is created within and disseminated from the Western paradigm, (practical) indigenous and local knowledge is often considered less valuable than modern scientific epistemologies (Briggs, 2005). For example, a traditional amaXhosa rural resident in the Eastern Cape province (South Africa) may share their local construction techniques. Accordingly, they built what is known as a *rondavel* in which to live. It has cylindrical walls made of cow dung and a conical roof made of spear

[10] If being a human is—at least to some degree and in most cases—marked by the ability to produce and share knowledge, and if some knowers are (still) seen as epistemically lesser in this postcolonial context, then we can say that such knowers who are considered inferior are being dehumanised. An injustice occurs, which becomes 'prima facie problematic from a moral point of view, while it implies domination, disrespect, a sense of inferiority and erosion of dominated people's epistemic framework, collective identity, and existential security' (Collste, 2019: 162).

[11] For example, it should be noted that 'indigenous knowledge and local adaptive responses' is mentioned only once in the National Climate Change Response White Paper (South African Department of Environmental Affairs, 2011: 23).

grass. This traditional structure can help mitigate the impact of severe weather better than more modern construction techniques using brick and cement to build more angular constrictions (see Apraku et al., 2018: 15, Figure 3). Such 'non-scientific' and 'primitive' socio-epistemologies are deprioritised in relation to technological and scientific solutions regarding climate change knowledge (Briggs, 2005).[12] This stands in contrast to the IPCC's (2023) acknowledgement that indigenous and local knowledge can also contribute to the dialogue on the formulation and creation of climate change mitigation and adaptation strategies and policies.

Epistemic *Othering*

When we consider that many national governments often use the IPCC reports to make decisions on climate change strategies and policies to be implemented, it becomes obvious that the dominant climate change knowledge being disseminated in a developing country like South Africa is based on information created and received from developed countries (see Pasgaard et al., 2015 for a detailed discussion).[13] This techno-scientific knowledge in the centre of the climate change episteme is firmly situated in the Western paradigm. Here, indigenous and local

[12] An example of African indigenous knowledge is ubuntu, which plays a significant role in climate change adaptation by fostering collective action, community resilience and sustainable practices within local (sub-Saharan) contexts (see Apruka et al., 2018; Terblanché-Greeff, 2019). This cultural focus on community and its implications for climate change adaptation is supported by Shingirai S. Mugambiwa (2018: 3), who states that '[c]ommunity-based adaptation is helpful in reducing the vulnerability as well as improving the resilience of the local people to climatic variability and change'.

[13] Homogenising all developing countries of the Global South is not a useful practice. As such, it must be noted that the Government of Botswana takes a different stance regarding the (over)reliance on Western science and the acknowledged importance and implementation of indigenous and local knowledge as posed in their National Adaptation Plan (NAP) (see Government of Botswana, 2020). 'For the communities to actively participate in the NAP process, it is critical that their customs, practices, and Indigenous and Traditional Knowledge are utilized *in addition* to technical or scientific approaches. This participation will reinforce that they are equal partners and stakeholders in decision-making.' (Government of Botswana, 2020: 14, emphasis added). This shows that although *some* Global South governments (like South Africa) *predominantly* focus on Western scientific knowledge when developing their climate change policies, it is not the case for *all* developing countries in the Global South.

knowledge is pushed to the periphery.[14] This can create epistemic *othering* through self-other distantiation.

Lajos Brons (2015: 70) describes the process of *othering* through self-other distantiation as follows:

> Othering is the simultaneous construction of the self or in-group and the other or out-group in mutual and unequal opposition through [the] identification of some desirable characteristic that the self/in-group has and the other/out-group lacks and/or some undesirable characteristic that the other/out-group has and the self/in-group lacks. Othering thus sets up a superior self/in-group in contrast to an inferior other/out-group, but this superiority/inferiority is nearly always left implicit.

Brons (2015: 72) elaborates when he states that

> … the conclusion of othering is self-other distantiating and dehumanizes the other, but this does not necessarily have to take the form of an affirmation of self-superiority and other-inferiority. Although othering often sets up a superior self/in-group in contrast to an inferior other/out-group, it can also create distance between self/in-group and other/out-group by means of a dehumanizing over-inflation of otherness. The other then, is not so much (implicitly) inferior, but radically alien. In either case, the effect is a near impenetrable border between the self/in-group and the inferior and/or radically alien other/out-group, "justifying" social exclusion, discrimination, and/or subjection.

On the basis of the preceding, it is argued that self-other distantiation resulting in an inferior other/outgroup and/or radical alienisation of the other/outgroup regarding differing epistemologies relate to epistemic oppression and wilful hermeneutical ignorance in epistemic injustice. There is a so-called barrier between the epistemic resources of the hermeneutically advantaged and disadvantaged groups. This epistemic *othering* becomes prevalent when describing othered epistemologies using dichotomies like North/South, scientific/indigenous and centre/periphery in the dominant paradigm of the Global North.

This epistemic *othering* can severely affect the hermeneutically marginalised South African populace when it comes to climate change knowledge. The resultant wilful hermeneutical ignorance regarding alternative epistemic sources and/or the devaluation of indigenous and local knowledge on climate change can lead to

[14] Although it is mentioned that such knowledge is of value, it is often not seriously considered, which is predominantly the case in the official South African climate change documentation. This is done despite research showing that indigenous and local knowledge can contribute to climate change mitigation and adaptation in the South African context (see Roos et al., 2010; Apraku et al., 2018; Kom et al., 2023).

ineffective or inappropriate climate change mitigation and adaptation strategies, as they may fail to account for local contexts, community needs and cultural practices (Byskov and Hyams, 2022).[15]

A further consequence of developing climate change mitigation and adaptation strategies and policies based solely on the Western paradigm is the potential for *epistemic violence*. The root of epistemic violence can be found in *othering* (Pérez, 2019). This form of violence encompasses a range of practices and mechanisms through which knowledge is devalued, distorted, or rendered inaccessible, often in ways that perpetuate social hierarchies and inequalities (Pérez, 2019). Accordingly, epistemic violence often includes subtle forms of oppression, such as the delegitimisation of indigenous and local knowledge systems (that differ from modern scientific epistemology) and the imposition of dominant narratives. Dotson (2011: 236) elaborates that '[a]n epistemic side of colonialism is the devastating effect of the "disappearing" of knowledge, where local or provincial knowledge is dismissed due to privileging alternative, often Western, epistemic practices'. In the postcolonial context, the dominant narrative pertains to the techno-scientific climate change knowledge created and disseminated by the developed countries of the Global North to be utilised by all, including developing countries in the Global South. The preceding is indicative of the remnants of coloniality that marginalised South Africans still experience.

The Way Forward

Considering that the lack of recognition through the social subordination of the *other* is a prominent aspect of epistemic injustice, we can argue that recognition is needed for epistemic justice (Collste, 2019). That is to say, everyone—regardless of their level of *perceived* hermeneutic power—should be equally included and acknowledged as knowers in prevalent dialogues (such as the discursive practices in political debate, public reasoning and, ultimately, decision-making) to ensure epistemic justice. However, the mere inclusion of previously marginalised groups is insufficient to facilitate epistemic justice. We should also challenge the underlying structural-identity prejudices and

[15] Additionally, for strategies to be effective, the South African national government needs buy-in from the public. If the climate change strategies are based on the Western techno-cratic mould—instead of being context-specific—the local communities may be unable to identify with the strategies or be unwilling to take ownership of the implementation (Leal Filho et al., 2022), thus contributing to the inefficacy of the strategies.

resultant systemic structures of inequalities preserving epistemic injustice, such as the socio-economic and linguistic inequalities prevalent in South Africa (to name but two). That is, the decolonisation of dominant epistemes is required to address the persistent coloniality of current epistemologies better (Cummings et al., 2023). Such decolonisation of knowledge systems requires '[...] processes and actions that intentionally dismantle the entrenched, unequal patterns of knowledge creation and use that emanate from our colonial past' (Cummings et al., 2021: 65).

Lerato Posholi (2020: 285) states that 'decolonizing knowledge involves working with Eurocentric concepts and categories, subjecting them to critical scrutiny and reconstructing them in light of the experiences and perspectives of marginalized groups to then use them as tools to make sense of our world'. The aim is not to replace modern scientific epistemology with indigenous and local knowledge. Instead, when it comes to climate change knowledge, there should be a break away from the North/South, scientific/indigenous and centre/periphery dichotomies through decolonisation. In this way, the dominant epistemology can be decentred without being replaced. It calls for an *acentrism*, where 'no single group can claim a center stage except in the context of incomplete descriptions. Each group must constitute its own center, but these polycenters must coalesce to form the whole' (Verharen, 1995: 66). Similarly, Boaventura de Sousa Santos and colleagues call for an *ecology of knowledges*, which is 'an invitation to the promotion of non-relativistic dialogues among knowledges, granting "equality of opportunities" to the different kinds of knowledge engaged in ever broader epistemological disputes aimed both at maximizing their respective contributions to build a more democratic and just society and at decolonizing knowledge and power' (de Sousa Santos et al., 2008: xx). Ideally, a complementary and reciprocal relationship between (once opposing) epistemologies can exist. There is a need for expanding epistemic resources paired with dismantling 'the apartheid of theoretical domains through renewed forms of critique and epistemic creolization' (Maldonado-Torres, 2008: 7).

Turning his gaze specifically to Africa, James O. Ogone (2017: 16) continues by stating that '[e]pistemic justice has [the] potential for granting African knowledge a fair opportunity to be heard alongside the dominant epistemologies. It is thus a moral responsibility owed to the less powerful knowledges by the dominant epistemologies [of developed countries in the Global North] [...] in the realization of democratic ideals integral to the definition of modern societies'. The onus of establishing epistemic justice—and climate change epistemic justice—can thus be laid at the feet of a democratic government like that of South Africa, with its overreliance on climate change knowledge based on the Western paradigm.

The preceding shows that recognition, inclusive discourse and the decolonisation of epistemologies are surely needed to address epistemic injustice. Still, we can go further and state that to address the two forms of epistemic injustice discussed, a government (like that of South Africa) has a twofold moral responsibility. That is to create the circumstances for the wilful uptake of indigenous epistemic resources into the dominant resource pool and to include indigenous and local knowledge in climate change mitigation and adaptation strategies and policy processes, thus facilitating epistemic justice in the context of climate change knowledge.

It is not sufficient to rely solely on the climate change knowledge disseminated by the developed countries of the Global North to make important context-specific and nuanced decisions about climate change mitigation and adaptation strategies and policies. It requires an unlearning of an overreliance on information based on the Western paradigm and the enabling of the fair treatment of groups in the process of interpretation and knowledge acquisition. The latter can be facilitated through an Afrocentric approach to climate change knowledge. This approach 'provides methods and practices that African people can use for making sense of their everyday experiences from an indigenous African's point of view' (Mugambiwa, 2018: 4). Such sense-making of reality can contribute to the overall social hermeneutical resource pool in South Africa, which is currently marked by Western epistemic dominance. Unlearning and decolonisation can lead to epistemic justice, which accentuates the need for recognising and respecting diverse experiences, perspectives, and forms of knowledge, especially in contexts where power imbalances or systemic injustices exist.

Conclusion

Climate change knowledge is often created and disseminated unidirectionally from the developed countries of the Global North to the developing countries of the Global South. For example, this is done by distributing scientific knowledge such as the IPCC's assessment reports, which are considered authoritative in informing national governmental decision-making on climate change mitigation and adaptation strategies and policies. This brings into question who the players are in the aforementioned creation and distribution of climate change knowledge. It becomes clear that everyone is not given an equal voice (as preference is given to certain dominant epistemologies), which leads to epistemic injustice. Furthermore, it is said that '[e]pistemic injustice comprises unfair treatment in knowledge-related and communicative practices in which the voices, experiences and problems of marginalized individuals, communities and societies are not being taken seriously' (Cummings et al., 2023: 1965).

It has been argued that epistemic injustice (be it hermeneutical or contributory) is endemic to some developing countries of the Global South—such as South Africa—when it comes to climate change knowledge. Here, hermeneutically marginalised communities (specifically, groups with indigenous and local knowledge) often find their experiential knowledge dismissed, misrecognised, or discounted. To address this issue will require more than merely amplifying the once-silenced voices. Instead, it requires the decolonisation of climate change knowledge. That is, we must challenge the underlying structural-identity prejudices that result in inequalities that perpetuate epistemic injustice. By doing so, we can foster equal and inclusive discursive practices concerning climate change knowledge and the relevant, diverse epistemologies to be included in such knowledge creation and dissemination.

References

Apraku, A., Akpan, W., & Moyo, P. (2018). Indigenous knowledge, global ignorance? Insights from an Eastern Cape climate change study. *South African Review of Sociology*, 49, 1–21. https://doi.org/10.1080/21528586.2018.1532813

Bhargava, R. (2013). Overcoming the epistemic injustice of colonialism. *Global Policy*, 4(4), 413–417. https://doi.org/10.1111/1758-5899.12093

Briggs, J. (2005). The use of indigenous knowledge in development: Problems and challenges. *Progress in Development Studies*, 5(2), 99–114

Brons, L. (2015). Othering, an analysis. *Transcience. A Journal of Global Studies*, 6(1), 69–90

Byskov, M. F., & Hyams, K. (2022). Epistemic injustice in climate adaptation. *Ethical Theory and Moral Practice*, 25(4), 613–634. https://doi.org/10.1007/s10677-022-10301-z

Collste, G. (2019). Cultural pluralism and epistemic injustice. *Journal of Nationalism, Memory & Language Politics*, 13(2), 152–163. https://doi.org/10.2478/jnmlp-2019-0008

Cummings, S. J. R., Munthali, N., & Shapland, P. (2021). A systemic approach to the decolonisation of knowledge. In D. Ludwig, B. Boogaard, P. Macnaghten, & C. Leeuwis (Eds.), *The politics of knowledge in inclusive development and innovation* (pp. 65–79). Sussex: Routledge. https://doi.org/10.4324/9781003112525

Cummings, S., Dhewa, C., Kemboi, G., & Young, S. (2023). Doing epistemic justice in sustainable development: Applying the philosophical concept of epistemic injustice to the real world. *Sustainable Development*, 31(3), 1965–1977. https://doi.org/10.1002/sd.2497

de Sousa Santos, B., Nunes, J. A., & Meneses, M. P. (2008). Introduction. In B. de Sousa Santos (Ed.), *Another Knowledge Is Possible: Beyond Northern Epistemologies*. Verso, London.

Dotson, K. (2011). Tracking epistemic violence, tracking practices of silencing. *Hypatia*, 26(2), 237–257. https://doi.org/10.1111/j.1527-2001.2011.01177.x

Dotson, K. (2012). A cautionary tale: On limiting epistemic oppression. *Frontiers: A Journal of Women Studies*, 33(1), 24–47. https://doi.org/10.5250/fronjwomestud.33.1.0024

Dotson, K. (2014). Conceptualizing epistemic oppression. *Social Epistemology*, 28(2), 115–138. https://doi.org/10.1080/02691728.2013.782585

Fricker, M. (1999). Epistemic oppression and epistemic privilege. *Canadian Journal of Philosophy Supplementary*, 25(25), 191–210

Fricker, M. (2006). Powerlessness and social interpretation. *Episteme*, 3(1–2), 96–108. https://doi.org/10.3366/epi.2006.3.1-2.96

Fricker, M. (2007). *Epistemic injustice: Power & the ethics of knowing*. Oxford: Oxford University Press

Fricker, M. (2013). Epistemic justice as a condition of political freedom? *Synthese*, 190(7), 1317–1332. https://doi.org/10.1007/s11229-012-0227-3

Fricker, M. (2017). Evolving concepts of epistemic injustice. In I. J. Kidd, J. Medina, & G. Pohlhaus Jr. (Eds.), *The Routledge handbook of epistemic injustice* (pp. 53–60). Routledge

Government of Botswana (2020). *National Adaptation Plan Framework for Botswana*. Retrieved January 5, 2024, from https://napglobalnetwork.org/wp-content/uploads/2020/06/napgn-en-2020-nap-framework-for-botswana.pdf

Hofmeyr, J. (2024). Summary of results: *Afrobarometer Round 9 survey in South Africa, 2022*. Compiled by the Institute for Justice and Reconciliation. Retrieved September 11, 2024, from https://www.afrobarometer.org/wp-content/uploads/2024/03/Summary-of-results-South-Africa-Afrobarometer-Round-9-7mar24.pdf

Holroyd, J., & Puddifoot, K. (2020). Epistemic injustice and implicit bias. In E. Beeghly, & A. Madva (Eds.), *An Introduction to implicit bias: Knowledge, justice, and the social mind*. Routledge. https://doi.org/10.4324/9781315107615-6

IPCC (2023). Summary for policymakers. In *Climate Change 2022 – Impacts, adaptation and vulnerability: Working Group II contribution to the Sixth Assessment Report of the Intergovernmental Panel on Climate Change* (pp. 3–33). Cambridge University Press. https://doi.org/10.1017/9781009325844.001

Jasanoff, S. (2004). *States of knowledge: The co-production of science and the social order*. Routledge. https://doi.org/10.4324/9780203413845

Karlsson, S., Srebotnjak, T., & Gonzales, P. (2007). Understanding the North-South knowledge divide and its implications for policy: A quantitative analysis of the generation of scientific knowledge in the environmental sciences. *Environmental Science & Policy*, 10(7–8), 668–684. https://doi.org/10.1016/j.envsci.2007.04.001

Kidd, I. J., Medina, J., & Pohlhaus, G. Jr., (2017). Introduction to the Routledge handbook of epistemic injustice. In I. J. Kidd, J. Medina, & G. Pohlhaus Jr. (Eds.), *The Routledge handbook of epistemic injustice* (pp. 1–9). Routledge

Knorr-Cetina, K. D. (1999). *Epistemic cultures: How the sciences make knowledge*. Harvard University Press

Kom, Z., Nethengwe, N. S., Mpandeli, S., & Chikoore, H. (2023). Indigenous knowledge indicators employed by farmers for adaptation to climate change in rural South Africa. *Journal of Environmental Planning and Management*, 66(13), 2778–2793. https://doi.org/10.1080/09640568.2022.2086854

Leal Filho, W., Barbir, J., Gwenzi, J., Ayal, D., Simpson, N. P., Adeleke, L., Tilahun, B., Chirisa, I., Gbedemah, S. F., Nzengya, D. M., & Sharifi, A. (2022). The role of indigenous knowledge in climate change adaptation in Africa. *Environmental Science & Policy*, 136, 250–60. https://doi.org/10.1016/j.envsci.2022.06.004

Maldonado-Torres, N. (2007). On the coloniality of being: Contributions to the development of a concept. *Cultural Studies*, 21(2), 240–270. https://doi.org/10.1080/09502380601162548

Maldonado-Torres, N. (2008). *Against war: Views from the underside of modernity*. Durham: Duke University Press

Mugambiwa, S. S. (2018). Adaptation measures to sustain indigenous practices and the use of indigenous knowledge systems to adapt to climate change in Mutoko rural district of Zimbabwe. *Jàmbá. Journal of Disaster Risk Studies*, 10(1), 1–9. https://doi.org/10.4102/jamba.v10i1.388

National Planning Commission (2012). *National development plan 2030: Our future – Make it work*. Republic of South Africa. Retrieved January 5, 2024, from https://www.gov.za/sites/default/files/gcis_document/201409/ndp-2030-our-future-make-it-workr.pdf

Ogone, J. O. (2017). Epistemic injustice: African knowledge and scholarship in the global context. *Postcolonial Justice*, 191, 17–36. https://doi.org/10.1163/9789004335196_004

Pasgaard, M., Dalsgaard, B., Maruyama, P. K., Sandel, B., & Strange, N. (2015). Geographical imbalances and divides in the scientific production of climate change knowledge. *Global Environmental Change*, 35, 279–288. https://doi.org/10.1016/j.gloenvcha.2015.09.018

Pérez, M. (2019). Epistemic violence: Reflections between the invisible and the ignorable. *El lugar sin límites*, 1(1), 81–98

Pohlhaus, G. (2011). Relational knowing and epistemic injustice: Toward a theory of wilful hermeneutical ignorance. *Hypatia*, 1127(4), 715–735

Posholi, L. (2020). Epistemic decolonization as overcoming the hermeneutical injustice of Eurocentrism. *Philosophical Papers*, 49(2), 279–304. https://doi.org/10.1080/05568641.2020.1779604

Republic of South Africa. (2022). *Climate Change Bill B9-2022*

Rheinberger, H. J. (2010). *On historicizing epistemology: An essay*. Stanford University Press

Roos, V., Chigeza, S., & Van Niekerk, D. (2010). Coping with drought: Indigenous knowledge application in rural South Africa. *Indilinga African Journal of Indigenous Knowledge Systems*, 9(1), 1–11

Rosenzweig, C., Casassa, G., Karoly, D. J., Imeson, A., Liu, C., Menzel, A., Rawlins, S., Root, T. L., Seguin, B., & Tryjanowski, P. (2007). Assessment of observed changes and responses in natural and managed systems. In M. L. Parry, O. F. Canziani, J. P. Palutikof, P. J. van der Linden, & C. E. Hanson (Eds.), *Climate Change 2007: Impacts, Adaptation and Vulnerability. Contribution of Working Group II to the Fourth*

Assessment Report of the Intergovernmental Panel on Climate Change (pp. 79–131). Cambridge, UK: Cambridge University Press

South African Department of Environmental Affairs (2011). *National climate change response white paper*. Republic of South Africa. Retrieved January 5, 2024, from https://www.gov.za/sites/default/files/gcis_document/201409/nationalclimatechangeresponsewhitepaper0.pdf

Sulla, V., Zikhali, P., & Cuevas, P. F. (2022). *Inequality in Southern Africa: An assessment of the Southern African Customs Union (English)*. Washington, DC: World Bank Group. Retrieved January 5, 2024, from http://documents.worldbank.org/curated/en/099125303072236903/P1649270c02a1f06b0a3ae02e57eadd7a82

Terblanché-Greeff, A.C. (2019). Ubuntu and environmental ethics: The West can learn from Africa when faced with climate change. In M. Chemhuru (Ed.). *African environmental ethics: A critical reader* (pp. 93–109). Switzerland: Springer

Tobi, A. (2022). Epistemic injustice and colonisation. *South African Journal of Philosophy*, 41(4), 337–346 https://doi.org/10.1080/02580136.2023.2199605

Trisos, C. H., Adelekan, I. O., Totin, E., Ayanlade, A., Efitre, J., Gemeda, A., Kalaba, K., Lennard, C., Masao, C., Mgaya, Y., Ngaruiya, G., Olago, D., Simpson, N. P., & Zakieldeen, S. (2022). Africa. In H. O. Pörtner, D. C. Roberts, M. Tignor, E. S. Poloczanska, K. Mintenbeck, A. Alegría, M. Craig, S. Langsdorf, S. Löschke, V. Möller, A. Okem, & B. Rama (Eds.), *Climate Change 2022: Impacts, Adaptation and Vulnerability. Contribution of Working Group II to the Sixth Assessment Report of the Intergovernmental Panel on Climate Change* (pp. 1285–1455). Cambridge & New York, NY: Cambridge University Press. https://doi.org/10.1017/9781009325844

UNESCO (2021). *Local and indigenous knowledge systems (LINKS)*. Retrieved January 4, 2024, from https://en.unesco.org/links

Verharen, C. C. (1995). Afrocentrism and acentrism: A marriage of science and philosophy. *Journal of Black Studies*, 26(1), 62–76

Wa Thiong'o, N. (1987). *Decolonising the mind: The politics of language in African literature*. Harare: Zimbabwe Publishing House

Zuma-Netshiukhwi, G., Stigter, K., & Walker, S. (2013). Use of traditional weather/climate knowledge by farmers in the South-Western Free State of South Africa: Agrometeorological learning by scientists. *Atmosphere*, 4(4), 383–410. https://doi.org/10.3390/atmos4040383

Participatory Governance in Botswana's Climate Change Policy: The Case of the *Kgotla*

6

Elisabeth Alm

Abstract

Botswana's government has positioned itself as an active participant in climate governance through international, regional, and national commitments. However, the effectiveness of these commitments at the local level remains unclear. This chapter examines Botswana's policy positioning on participatory governance in its climate change response, with a focus on the role of the *kgotla*, an indigenous leadership forum. While the *kgotla* is formally recognised as a space for local engagement and the integration of indigenous knowledge, its actual influence in climate policy decision-making is limited. The government promotes decentralisation in theory but retains central control, often using the *kgotla* as a symbolic tool rather than a meaningful platform for local participation. Furthermore, while national policies emphasise the inclusion of indigenous knowledge, mechanisms for its implementation remain vague. The chapter argues that Botswana's climate governance reflects a gap between policy rhetoric and practice, where local engagement is acknowledged but not effectively realised. This chapter highlights the contradictions within Botswana's dual governance system and raises critical questions about the future role of the *kgotla* in shaping truly participatory climate governance while also recognising the ambiguous character of the *kgotla* itself.

E. Alm (✉)
University of Freiburg, Freiburg, Germany
e-mail: Elisabeth-alm@gmx.de

A. C. Terblanché-Greeff et al. (eds.), *Climate Change and the Postcolonial*, Kulturelle Figurationen: Artefakte, Praktiken, Fiktionen,
https://doi.org/10.1007/978-3-658-49328-8_6

Keywords

Indigenous leadership · Climate change governance · Postcolonialism · Local climate action

Introduction

The Republic of Botswana (hereafter referred to as Botswana) is a landlocked country in Southern Africa bordering South Africa, Namibia, Zambia and Zimbabwe. Botswana has a semi-arid climate, with the vast Kalahari Desert covering most of the country. The northern parts of the country, however, are mainly made up of the Okavango Delta and the Zambezi River. Botswana is highly vulnerable to climate change (Dube, 2003). An increase in extreme floods and droughts in recent years has led to rising levels of food insecurity as the livelihood of many Botswana citizens is threatened. The majority of citizens are involved in subsistence livestock and horticultural production; both of these agricultural activities are strongly impacted by climate change (Mugari et al., 2020). Another climate change impact is water scarcity affecting rural and urban areas alike, risking the livelihoods of the local people (Mpandeli et al., 2018). Besides agricultural production, nature-based tourism is another source of income for parts of the population, especially in the northern parts of Botswana, near the Okavango Delta. This, too, is threatened by climate change as the biodiversity in the ecosystem is decreasing, making it less attractive for tourists to visit (Swemmer et al., 2018). Furthermore, climate change has led to an increase in viral infections and diseases, such as diarrheal disease, in the country, with a healthcare system not equipped to treat the increasing number of patients (Alexander et al., 2013). Due to the stark impact of climate change on Botswana, the government is enacting climate change governance at multiple levels.

As a signatory of the Paris Agreement, the government of Botswana shows its commitment to lowering greenhouse gas emissions internationally. The *Botswana Vision 2036* policy, with its third pillar focusing on sustainable development and growth, was unveiled just weeks before the Paris Agreement came into effect in 2016. This policy aligns with the content of the Paris Agreement, further showcasing Botswana's interest in complying with international measures (Mogomotsi et al., 2018). In accordance with the Paris Agreement, Botswana's government submitted its intended nationally determined contribution to the United Nations Framework Convention on Climate Change (UNFCCC) in 2016 and its *Third*

National Communication in 2019 (Botswana Government, 2016; Ministry of Environment, Natural Resources, Conservation and Tourism [MoENRCT], 2019). Furthermore, Botswana delegates regularly attend the annual United Nations Conference of Parties (COP) to represent their interests in an international forum.

Regionally, Botswana is involved in climate change governance through the Southern African Development Community (SADC), which has its seat in Gaborone, the capital city of Botswana. Together with 15 other countries in Southern Africa, Botswana has developed regional climate protection policies such as the *Regional Green Economy Strategy for Sustainable Development and Action Plan* (SADC, 2015b), the *Regional Agriculture Policy*, the *SADC Science, Technology and Innovation Climate Change Response Framework* (SADC, 2015a), and the *Protocol on Environmental Management for Sustainable Development* (SADC, 2014). Furthermore, in 2015, the SADC published its *SADC Climate Change Strategy and Action Plan*, which laid out a comprehensive framework for implementing climate change adaptation and mitigation measures (SADC, 2015c).

On a national level, the government of Botswana produced its first national strategy on climate change in 2018, relatively late compared to other countries, both on a regional level and a broader international scale. Nevertheless, *A National Climate Change Strategy for Botswana* document was published in December 2018 focusing on adaptation, mitigation and cross-cutting themes such as gender and communication and knowledge management (UNDP, 2018). This is complemented by *The Botswana Climate Change Policy*, which was implemented in 2021 on the basis of the principles of the national strategy (MoENRCT, 2021).

Scientific discourse has repeatedly highlighted the importance of localising climate action over the past decade (Bond, 2010). Greenhouse gas emissions occur in specific places, and so should any practices aimed at reducing emissions (Bond, 2010). Furthermore, criticism towards international, regional, or even national adaptation measures often highlights that important factors such as geographical, political, and social conditions differ between regions, nations and even between local places (Bond, 2010). Consequently, national or even broader policies are considered too standardised to adequately fulfil the needs of specific communities (Brown et al., 2021). This is illustrated by the many policies and agreements listed above by the SADC. While they show Botswana's general commitment towards climate action, they are made to represent 15 different countries, each with various cultures, languages, and overall diverse contexts. The holistic approach of these policies does not allow for the situated needs of individual communities to be represented. Including local communities in policy formation, for example, through the *kgotla* (an indigenous leadership forum), can lead to positive innovation within climate change governance (Bond, 2010). Meghan Bond (2010) identifies certain prerequisites as ideal for local participation in climate change governance. Such particulars include having a designated space for communities to voice their input and concerns, receive information or participate in workshops (Bond, 2010).

Community spaces fostering dialogue are a feature of local communities in many different cultures. In Botswana, the *kgotla* represents an indigenous leadership system, which has been a part of the country's culture and political structure since pre-colonial times (Obasi and Lekorwe, 2014). Its primary function is to serve as an open forum for the community to discuss issues that are important to them. The *kgotla* is led by the *kgosi* (chief), who, in turn, uses the forum to disseminate information to the community. It is, then, a direct exemplar of the designated space described by Bond (2010). Furthermore, the *kgotla* and its role in society is institutionalised within the Botswana constitution through the *Bogosi Act* of 2008 (Government of Botswana, 2008), which replaced the *Chieftainship Act* of 1987. This local, indigenous leadership system with solid ties and legitimisation in the community, therefore, equips the Botswana government with the opportunity to enact meaningful localised climate change governance. Nevertheless, within this postcolonial context, the incorporation of indigenous leadership in governance is confronted with specific challenges. For example, there is a devaluing of the expertise and knowledge collected and represented by the *kgotla*. The knowledge that is (re-)produced at the *kgotla* meetings can be described as indigenous. That is,

> knowledge that is fixed in the social history of individuals or communities; this knowledge includes people's historical development as well as their social and economic disposition, and the logical and innovative nature of their personality as a group. [...] [I]t is firmly linked to subsistence practices, and it is also context, location and culture specific, usually transmitted orally and informally. (Seleka, 2018: 24)

Within the postcolonial project of modernity, traditional practices lose their legitimacy (Ciaffa, 2008), extending to indigenous knowledge and indigenous leadership structures. With the exception of South Africa, Zimbabwe, and Namibia (which was under South African rule until 1990), the other Southern African countries gained independence from their European colonisers in the 1960s onwards. During the formation of their postcolonial political systems, the newly independent countries often followed the democratic structures of already established European nations. They modelled their state in their colonisers' image, following a framework of modernity with the goal of progress at its centre, rendering traditional practices superfluous. This ties into the concept of wilful hermeneutic ignorance (as explored by Terblanché-Greeff in this volume). Following Gaile Pohlhaus' definition (2012) in the Botswana context, wilful hermeneutic ignorance occurs when marginalised groups, such as rural communities, communicate using their epistemic resources, like indigenous knowledge.[1] Here, the dominant sphere of discourse and governance does not recognise their epistemic system as valid and continues to exclude it from their discourses. The marginalised communities are aware of this exclusion of their indigenous knowledge and are trying to change it. In Botswana, the role of the *kgotla* is even institutionalised in order to protect the institution. Despite these efforts, the dominant group wilfully chooses not to take knowledge produced as a side product by the *kgotla* into consideration when making sense of phenomena such as climate change. Hence, local climate change governance in Botswana faces the duality of having a decentralised political system with the *kgotla*, a space often described as well-equipped to take part in local governance, while simultaneously operating in a postcolonial world order that still perpetuates wilful hermeneutic ignorance and therefore undervalues the expertise offered by the *kgotla*.

The aim of this chapter is to understand the Botswana government's policy positioning on the integration of participatory governance in the form of the *kgotla* in its climate change response. Including the perspective of the *kgotlas* themselves is rather challenging, as no formalised records are available to the public. While the literature suggests that the *kgotlas* enjoy high levels of legitimacy and support in the country—even though this has been slowly declining in recent years (Ciaffa, 2008)—there is no information on the *kgotlas'* approach to climate change governance. This is further highlighted by the fact that the *kgotlas* represent the local, village level in Botswana and that *kgotlas* in diverse regions of the country (made up of various tribes present in Botswana) may have differing approaches to climate change and prioritise climate action differently. Hence, this chapter will mostly focus on the integration of participatory governance from the perspective of the national government.

[1] The understanding of wilful hermeneutic ignorance presented in this chapter follows a simplistic approach to the concept. Due to the nature of the chapter, a more in-depth analysis of wilful hermeneutic ignorance is not possible. It is, however, important to note that the case of Botswana is more complex, especially due to the institutionalisation of the *kgotla* within the political system of Botswana, lending it an unusual level of legitimacy in regard to discussing wilful hermeneutic ignorance. Furthermore, the main objective of the *kgotla* within the community is not climate communication and knowledge production.

This chapter will first outline the twofold political system of Botswana, constituted by the modern system and the *kgotla* system, and the interaction between the two. Following this section will be an overview of Botswana's climate change governance and the role the *kgotla* plays in localising it and centring community needs before posing the conclusion.

Botswana's Government System

To understand climate change governance within Botswana, the political system of the country needs to be introduced. This consists of two juxtaposed parts: the democratically elected, modern system implemented after Botswana gained independence from the United Kingdom in 1966, and the traditional indigenous leadership system, which was already in place during pre-colonial times. Within a postcolonial context, both parts of the political system are vying to achieve and maintain a certain level of influence and establish their authority, albeit on different levels within the state. First, there will be an overview of the modern system followed by an explanation of the indigenous leadership system of Botswana, as well as the interaction between the two.

Modern System

Since the country gained independence, Botswana has adopted a republican system of government. The executive branch is headed by the President, who is elected by the National Assembly (Obasi and Lekorwe, 2014). The National Assembly, also called the House of Parliament, is led by the President and makes up the legislative branch (Obasi and Lekorwe, 2014). The *Ntlo ya Dikgosi* (House of Chiefs), the representatives of the indigenous leadership structure, has a consultative function to the National Assembly and vets the created laws and policies to uphold the values and traditions of the citizens of Botswana (Tlou, 2021). Lastly, Botswana has an independent judiciary institution handling the country's high-profile criminal and civil cases following Roman Dutch law (Tlou, 2021).

Issac Obasi and Mogopodi Lekorwe (2014) describe the different levels of Botswana's governmental system as follows. The country is divided into different districts and sub-districts, and then into villages. On the district level, there are four distinct political structures, viz., the District Administration, District Council, Land Boards, and Tribal Administration.

The District Administration is headed by the District Commissioner, appointed by the President, and is responsible for coordinating local and central governmental activities in the district. Another structure is the District Council, whose members are democratically elected; it handles questions regarding primary education, health care, water supply and social and community services. The third structure is the Land Boards, composed of twelve members, some elected by the *kgotla* of the district and some appointed by the Ministry of Land Management, Water and Sanitation Services. The Boards are responsible for controlling the allocation and use of land, which is decided upon by the *kgosi*. Lastly, there is the Tribal Administration, which consists of the *dikgosi* (chiefs, plural of *kgosi*) of the district. Its function is to act as a representative body for their local residents, settle disputes, and uphold customs and traditions.

The sub-districts follow a similar structure as the districts, except their functions are more localised. For example, every sub-district has a subordinate Land Board, whose role is to implement the district Land Board decisions, report back to it, and allow people in less populated areas to reach the Land Board without having to travel to the centre of the district (Government of Botswana, 2018). The sub-districts also have Sub-District Councils and Sub-District Administration Offices, which provide localised services of their district level counterparts. Furthermore, health facilities and education departments are located at sub-district level to allow fast access for local citizens.

Lastly, the Village Development Committee—along with the *kgotla*—operates on village-level in Botswana (Obasi and Lekorwe, 2014). It is a sub-committee of the *kgotla* and is elected by it. The Committee's role is to co-ordinate village development by raising funds, having consultations with both the *kgotla* and government officials and implementing small-scale village projects. This illustrates that the modern system is intertwined with the *kgotla* on the village level of governance.

This section explains the decentralisation of the Botswana government. On a national level, the government follows centralised structures. The decentralised nature of the political system comes into play at the district and local levels, where political responsibilities are outsourced to indigenous leadership entities, such as the *kgotla*. Therefore, Botswana has both a centralised national government as well as an overall decentralised political structure, creating an interactive hybridity.

Kgotla System

The *kgotla* is a 'formal public assembly associated with the institutions of traditional leadership' (Ngwenya and Kgathi, 2011: 250). In contrast to the modern system based on bureaucratic hierarchy, the *kgotla* is a system of 'familiar indigenous kinship' (Helle-Valle, 2002: 183). It is headed by the *kgosi*, whose succession is hereditary, creating politically influential families in the communities. The *kgotla* is an open forum (often held in an open space outdoors) for the community to gather (Helle-Valle, 2002). The *kgosi* uses this forum to disseminate information to the community and consult the community on government plans and policies on a national and local level (Molebatsi and Morobolo, 2021).

During pre-colonial times, the authority of the *kgosi* was unlimited and undefined (Sharma, 2005). In the time of colonialism, Botswana was not a colony of Great Britain but a protectorate. Hence, the colonial government of Great Britain allowed the tribal structures to stay intact and even institutionalised the powers of the *kgotla* (Sharma, 2005). For example, in 1899, Hut Tax was introduced, and the *dikgosi* were tasked with collecting the local taxes and, in return, were allowed to keep 10% of the revenue (Sharma, 2005).[2] After Botswana gained independence from colonial rule, many responsibilities and authoritative powers of the *kgotla* shifted towards the democratically elected centralised national government (Chieftainship Act, 1987). Nonetheless, this has not reduced the importance of the *kgotla* within the local communities.

One of the main principles of the *kgotla* is 'Mua lebe ubua la gagwe', meaning that everyone has the right to voice their opinion and to be heard (Tlou, 2021). At the *kgotla*, everyone can attend, and, officially, any adult can speak up and be heard (Helle-Valle, 2002). The *kgotla* is often praised for its participatory governance and system of direct democracy (Ngwenya and Kgathi, 2011). Even though decisions are made based on a majority rule principle, the main aim of discussions at the *kgotla* is to reach a consensus within the community regarding decisions or actions to be taken. Officially, finding common ground is possible as the *kgotla* is said to be independent from partisan politics and instead values people's individual opinions.

[2] The Hut Tax was a taxation system used by the British colonial administration in different African colonies and protectorates of the United Kingdom. In this system, each household was taxed simply for existing, rather than because they were accruing income or trading goods. This allowed the British administration to control local population sizes and also force people into colonial wage labour structures as many people had to work to pay the Hut Tax (Monageng, 2006).

However, while the decisions within the *kgotla* are based on consensus, the leadership of this assembly, the *kgosi*, is still based on a monarchical system. The *kgosi* enjoys a high privilege and authoritative status in the community and can technically freely make decisions without public consultation; nonetheless, the *kgosi* does mostly follow the will of the people (Obasi and Lekorwe, 2014). This can be explained by the communities' understanding of the role of the *kgosi*. Even though it is a system based on hereditary succession, the *kgosi* is a leader 'by the people' (Obasi and Lekorwe, 2014) meaning that while their chiefdom lineage provides divine legitimacy, they are supposed to act in the peoples' best interest (Obasi and Lekorwe, 2014). Even though the *Bogosi Act* outlines ways for a *kgosi* to be dismissed following customary laws (Government of Botswana, 2008), the cultural barriers to doing so are high. Moreover, not valuing the *kgosi's* opinion is seen as highly disrespectful, and, consequently, people will oftentimes not speak out against them (Phago and Molosi-France, 2018). The same can be said for the opinions of those with close personal or familial ties to the *kgosi* (Phago and Molosi-France, 2018). The reality of this cultural dynamic directly contradicts the participatory and democratic nature that the *kgotla* is supposed to represent.

Moreover, during pre-colonial and colonial times, the *kgotla* leadership (with its patriarchal notions) was not open to women. This has slowly changed since Botswana gained independence, and in 2003, the first female *kgosi*, Mosadi Semboko, came to power (Ngwenya and Kgathi, 2011). Today, there are officially no barriers stopping anyone from participating in the *kgotla*, but in reality, there are still clear hierarchies based on age, gender and tribe. Even how long someone can speak at the *kgotla* is regulated by their place in the hierarchy, with young women having the lowest rank (Helle-Valle, 2002). When asked in a study by Barbara Ntombi Ngwenya and Donald Kgathi (2011), the older men participating in the *kgotla* felt that the institution was very open and welcoming to all. On the other hand, in that same study, women, young people, people of low socio-economic status and ethnic minorities (due to tribalism within the country) felt they did not have equal opportunities to participate due to invisible cultural barriers (Ngwenya and Kgathi, 2011).[3] This means that when local decisions are made or when the government consults the *kgotla* during their policy formation

---

[3] Botswana has eight major tribes, which are also represented in the *Ntlo ya Dikgosi*. The concept of tribes as groups based on ethnic identity was created and implemented during colonialism as part of the divide-and-conquer tactic of the colonisers. Part of this tactic was to foster hostilities between different tribes. Even though today Botswana is one of the most tribally-integrated countries in Africa, discrimination due to belonging to a different tribe is still a regular occurrence and can affect a person's standing in society (Makgala et al., 2020).

process, the voices and inputs from already marginalised groups in society are, to a degree, excluded from these processes, which is reflected in the decisions and policies being implemented. While the *kgotla* is praised for its perceived participatory nature, this leads to the question of for whom precisely the institution is participatory and who is left watching from the sidelines.

Especially men and the elderly play an important role at the *kgotla*. At the end of the meetings, the elders usually summarise the contents of the session. These summaries oftentimes include many metaphors and traditional proverbs (Tlou, 2021). This form of communication, as well as the content of the deliberations at the *kgotla* represent indigenous knowledge. Therefore, during government consultations at the *kgotla*—as part of the official policy-making process—the communities offer indigenous knowledge in the form of feedback as their epistemic contribution to the government.

Following this description of the *kgotla* system it is important to highlight the problematic character of the *kgotla* as a political entity. Overall, the *kgotla* is a very contradictory body. On the one hand, it is officially based on consensus, but the *kgosi* has the final say and can choose to disregard the community's opinion. Furthermore, the *kgotla* is often praised for being participatory, but in reality, women, young people and people from a tribe different from the main one represented in that community are to a degree excluded from the decision-making process. So, while this chapter aims to analyse the role of the *kgotla* in Botswana's climate change governance, it bears pointing out that the *kgotla* itself is not an unflawed political entity.

Interaction Between the Systems

Within the modern government system, the *Ntlo ya Dikgosi* represents the *kgotla* system at the highest national level. It is officially a part of the legislative branch and is directly involved in policy creation for Botswana. Nevertheless, the *Ntlo ya Dikgosi* is in an advisory position, and consequently, the House of Parliament is not obligated to take its inputs into consideration (Ifezue, 2015). Hence, the involvement of the *dikgosi* on the national level can be seen as performative to legitimise the government's policies since the *dikgosi* enjoy a high level of support within the country.

As stated, the decentralised nature of the Botswana government system is especially characterised by the political system at the district and village level. According to Obasi and Lekorwe (2014), the District Administration acts as an intermediary between the village and national levels, highlighting the importance of village activities in Botswana. The District Council essentially forms the modern system on the district level and helps maintain access to education, healthcare and water for local residents (Obasi and Lekorwe, 2014). It is the Land Board and the Tribal Administration specifically that integrate the *kgotla* system into its structures. The Land Boards unite government-appointed and *kgotla*-elected representatives and function within a checks and balances system over the *dikgosi* in the district, who are in charge of access to land and other natural resources. Therefore, the Land Boards legitimise this role of the *dikgosi* while simultaneously limiting their power by acting as a corrective authority (Obasi and Lekorwe, 2014). Lastly, the Tribal Administration is similar to the *Ntla ya Dikgosi* on the national level, whereas it vows to uphold the local customs and traditions as well as perform the role of the *kgotla* on a district level, including the settling of disputes. Therefore, the Tribal Administration has both judicial and legislative power on the district level.

The Village Development Committees are the governmental acting arm of the *kgotla*. They are essentially responsible for implementing the decisions taken by the *kgotla*. Structurally, they are dependent on the *kgotla* because it elects their representatives. Especially on the village level, the modern system legitimises the *kgotla* by creating the Committees to increase the *kgotla's* capacity to implement their plans and remain financially independent from the government as the Committees are also involved in local fundraising activities.

Not only is the *kgotla* included in the modern system, but the government also plays a role in its operations, and government officials bring national issues to be discussed at the *kgotla*. This is part of their policy-creation process and is used to present progress reports (Obasi and Lekorwe, 2014). Yet, since the *kgotla* was originally meant to discuss local issues, which are more personal to the village community, the debates on national issues are often not as extensive and fruitful (Molebatsi and Morobolo, 2021). Furthermore, the professional jargon used in presenting the national issues creates a barrier between the government representatives and the people, turning the meetings into lectures rather than consultative sessions (Molebatsi and Morobolo, 2021). Accordingly, the *kgotla* can lose its participatory nature in national matters and has been accused of being a 'repository forum for government-orchestrated development agenda' (Ngwenya and Kgathi, 2011: 250).

One example of the government being accused of co-opting the *kgotla* for image purposes is the nationwide tour by the Minister of Presidential Affairs and Public Administration, Joram K. K. Kelebile, ahead of the national election in 2004 (Obasi and Lekorwe, 2014). He travelled through the country, consulting the village communities, answering questions and collecting information on local issues. While the media reported positively on this endeavour, the residents in the villages that were consulted complained that the minister barely answered any questions they had and overall felt the tour was useless as he could assume a new role after the election and would not be able to implement the wishes voiced by the communities (Obasi and Lekorwe, 2014). Minister Kelebile was accused of using the *kgotla* tour as a performative act for his election campaign, undermining the original purpose of the institution. More recently, in 2019, Member of Parliament for Tonota, Pono Moatlhodi, accused sitting President Mokgweetsi Masisi of pretending to tour *kgotlas* for national consultation but actually using the visits for his presidential campaign (Patriot, 2021). Moatlhodi claimed that after the election, the consultations conducted by President Masisi stopped, proving a misconduct and abuse of the *kgotla* system by President Masisi (Patriot, 2021).

Regardless, there have been instances of the government using the *kgotla* consultation to empower local voices as well. When an unpopular law is presented at the *kgotla*, the protest of the community can challenge such legislation (Molebatsi and Morobolo, 2021). For example, on a local level, a government proposal about land usage for physical development in Tlokweng were rejected, which led to a revision of the government proposal (Molebatsi and Morobolo, 2021). A national example is the rejection of the Constitutional (Amendment) Bill of 2024, which faced severe criticism from different civil society actors. The *Ntlo Ya Dikgosi* rejected the bill in their advisory function to the government, and the bill failed to pass in parliament (Mungure, 2024). Furthermore, in the past, when the Botswana government had budget constraints, they consulted the *kgotla* in different villages, asking the communities which projects they should prioritise with the limited funds they had available (Obasi and Lekorwe, 2014).

The role of the *kgosi* is now governmentally controlled, as the government needs to approve of the *kgosi* (Government of Botswana, 2008). According to the *Bogosi Act* of 2008, the criteria for approval from the Ministry of Local Government and Rural Development are kept vague, simply referring to whether the person is a right fit for leadership under the specific tribes' customary law (Government of Botswana, 2008). The *kgosi* is legally required to carry out the instructions given to them by the Minister of Local Government and Rural Development (Sharma, 2005). Failure to comply with the requests of the government can lead to the *kgosi* being suspended (Government of Botswana, 2008). Hence, the *Bogosi Act* gives the government a lot of influence over the *kgotla* system. Notably, there have been instances where traditional leadership, e.g., Kgosi Keneth Autlwetse, assumed the position of the Minister of Local Government and Rural Development, indicating that governmental positions of the dual political systems can be intertwined.

In summary, the overarching decentralised government system is a national priority in Botswana and is part of the constitution, as well as the *Second Transitional National Development Plan* of the country (National Planning Commission [NPC], 2023). This dual political structure helps empower rural communities in policy formulation and development projects (Ngwenya and Kgathi, 2011). Still, the relationship between the national-level government and the *kgotla* can be seen as performative, where the government uses the *kgotla* to legitimise its actions, and representatives of the *kgotla* system only have an advisory role rather than actual direct political influence. For example, the government has the financial power in the state and can, consequently, control the majority of decisions and projects implemented in the country. This also creates a dependency of the *kgotla* on the government (Phago and Molosi-France, 2018). The Village Development Committees have mechanisms for fundraising, but these are not comparable to the funds available to the central government, leaving less funding for local initiatives.

Botswana's Climate Change Governance

As previously discussed in the chapter, internationally, there is a call for the transformation of climate change governance using a localised, bottom-up approach (Bond, 2010). In theory, a decentralised political system, such as the one in Botswana, is ideal for implementing such an approach as local inputs are to be valued in policy formation, leading to better identification of long-term solutions to climate change-related problems. Especially in a country such as Botswana, which is very vulnerable to climate change (Dube, 2003), successful mitigation and adaptation policies can improve the quality of life for its citizens by increasing food security, ensuring access to healthcare and preserving people's way of life within agricultural production. This section first introduces the plans, reports, and policies of the Botswana government and the organisations it is a part of. This is followed by some examples of the implementation of policies aiming at local climate action.

For the analysis of Botswana government reports and policies, the focus is on the inclusion of the *kgotla* and indigenous knowledge. As previously explained, indigenous knowledge is a part of the *kgotla* as an institution. It is also regarded this way by the Botswana government, which is why, for the purpose of this chapter, there is a soft conflation of the two, the *kgotla* and indigenous knowledge. Furthermore, it is necessary to highlight that none of the documents discussed below actually offer a clear definition of what indigenous knowledge is.

Governmental Plans and Policies

Botswana's recognition of the importance of localised climate change governance is stated in the country's participation in the *SADC Climate Change Strategy and Action Plan* (SADC, 2015c). The regional-level plan divides adaptation measures into nine sectors, one being biodiversity. Within the biodiversity sector, it is the first principle to '[p]romote scientific and indigenous knowledge on the vulnerability of biodiversity to climate change' (SADC, 2015c: 5). Since indigenous knowledge is always context-specific, it indicates localisation of climate action. Yet, compared to other regional organisations' commitment to indigenous knowledge and governance in their communal reports and plans, SADC, of which Botswana is a part, lags behind. For example, the regional organisation Association of Southeast Asian Nations, in its latest *ASEAN State of Climate Change Report* (ASEAN Secretariat, 2021), explicitly states the high affectedness of indigenous communities by climate change and pushes for cross-sectoral inclusion and promotion of indigenous knowledge in climate action multiple times in its report. Even though signalling such a commitment does not necessarily translate into real-life localised policies, regionally, Botswana and SADC only minimally show such effort.

This trend continues on a national level. Earlier in this chapter, *The Botswana Climate Change Policy* (MoENRCT, 2021) and *A National Climate Change Strategy for Botswana* (UNDP, 2018) were introduced as national action plans. In the *National Climate Change Strategy for Botswana* there is a strategic intervention to '[e]nsure the incorporation and utilisation of indigenous knowledge systems in all relevant climate change activities' (UNDP, 2018: 105). The strategy goal is as follows:

> The government will also develop climate change research strategy as a cross sectoral mechanism that will ensure that climate change elements are added into formal, informal and non-formal sectors of education and training as a scarce skill. This strategy will assist in exploring and utilizing indigenous knowledge and technical expertise in decision making mechanisms. (UNDP, 2018: 105)

This strategy goal will be realised by including indigenous knowledge 'into all relevant climate planning documents, policies, and programmes' (UNDP, 2018: 105). The long-term target is to have indigenous knowledge incorporated and applied 'wherever appropriate'. This loose definition, however, provides the government with the freedom to interpret and apply it in whichever way they see fit. Therefore, while the inclusion of indigenous knowledge (through the *kgotla*) in decision-making processes is planned, there is no mechanism ensuring its implementation.

A second strategic intervention regarding indigenous knowledge implicates the *kgotla* in their function as a forum for dialogue:

> Community groups, youth groups, NGOs, civil society, and the private sector will be engaged regularly through climate change forums, town halls, and panel discussions to enhance the dialogue on climate change in Botswana with an added focus on learning from and supporting indigenous knowledge systems. (UNDP, 2018: 108)

The goal of this strategy is to '[e]nsure knowledge concerning climate change is accessible, well managed and communicated throughout all sectors' (UNDP, 2018: 108).

Based on this climate change strategy for Botswana (UNDP, 2018), *The Botswana Climate Change Policy* (MoENRCT, 2021) was formulated. In the policy, it is mentioned that community input was a part of creating this policy and was collected through governmentally facilitated consultation meetings in different *kgotlas* (MoENRCT, 2021). Hence, the localisation of climate change governance and initiatives is a well-documented goal of the Botswana government (MoENRCT, 2021; UNDP, 2018) and has already been implemented in some of their policy formation efforts.

However, the *Secondary Transitional National Development Plan* (NPC, 2023) of Botswana describes 'Challenges with Participatory Governance' (regarding traditional leadership). These include

> difficulties in identifying successors; *dikgosi* dabbling in the divisive enterprise of political activism in significant numbers, contrary to their traditional role as unifiers in their respective communities; and, skirmishes with the Government. (NPC, 2023: 141)

The NDP suggests that the *dikgosi* no longer represent the majority opinion of their communities but are rather pushing for their own political will to be implemented. Instances where this political will of individual *kgosi* goes against the laid-out plan of the government are described as 'skirmishes' by the report. This supports that the Botswana government does not appreciate the *dikgosi* voicing any disapproval of government policies.

Furthermore, according to the NDP document, the *kgotla* struggles with 'asserting authority over a westernised, youthful population' (NPC, 2023: 141). The points raised suggest a decline in political soft power within the *kgotla* system as influence over society decreases. Urbanisation processes in Botswana are drawing people towards the urban centres of the country (Statistics Botswana, 2014). The village-based *kgotlas* can, therefore, no longer rely on the younger generations to continue their traditions and legitimise their local influence.

The same document also identifies climate change as 'probably the major challenge facing the world'. (NPC, 2023: 18). However, when discussing mitigation and adaptation measures, the report takes a global perspective. It discusses global inequalities by saying the majority of mitigation efforts should lie with Global North countries, who are the leading global polluters (NPC, 2023: 18). According to the NDP, climate change is impacting the 'global economy' and is a 'global governance challenge'. Combatting the phenomenon is a 'global responsibility' and the '[d]evelopments at a national level will be influenced by national policy decisions, but also to a large extent, by global developments' (NPC, 2023: 19). The climate change chapter in the NDP does not reflect any sort of local approach to climate change governance but rather does the opposite. The NDP is a guiding document, which influences future policy initiatives and other government programs. This fact, combined with mentioning the *kgotla* in relation to challenges in the country, could suggest that the Botswana government may not implement its policy commitment to localise its climate change governance through the *kgotla* system regardless of its previously signalled intentions.

In the *National Adaptation Plan Framework for Botswana* (MoENRCT, 2020), the importance of vertical integration in climate change matters is highlighted, which links adaptation planning and implementation between different levels of government from the national level to the *kgotla*. This vertical integration is aimed at implementation activities and ensuring a 'consistent information flow from ministries to departments, districts, grassroots [which is the *kgotla*], and vice versa' (MoENRCT, 2020: 6). According to the Plan, this will be implemented through the establishment of District Climate Change Committees to link governance on different levels. Moreover, the Plan has a chapter on including indigenous and traditional knowledge in the adaptation process. In this context, indigenous knowledge encompasses 'customs and practices' such as 'traditional forest management practices that contribute to increased forest cover and land rehabilitation' (MoENRCT, 2020: 14). Since the *kgotla* itself is an indigenous practice, this chapter reinforces the intent of the government to utilise the *kgotla* in their climate change governance. Furthermore, including the *kgotla* in policy decisions regarding conservation measures such as forest management for land rehabilitation means involving the people directly in the protection of their immediate environment in which they practise agriculture to sustain themselves. The exact nature of this collaboration between the national government and local *kgotlas* is not described.

The *Botswana Climate Change Policy* (MoENRCT, 2021) briefly discusses issues regarding the *kgotla*. The *kgotla* itself is only mentioned in the background chapter for the policy itself, stating that consultations with the population were held at the *kgotla* to increase the legitimacy of the policy (MoENRCT, 2021). The policy also restates the inclusion of indigenous knowledge in forest management (MoENRCT, 2021). Other than those two mentions, the *kgotla* and indigenous knowledge are not represented. Throughout the policy, there is an insistence on integrating the 'local level' into the process of Botswana's climate change governance. However, as previously explained, this 'local level' in Botswana encompasses different types of institutions besides the *kgotla*, such as the Village Development Committees, meaning it is unclear what specific institution the policy refers to. Sometimes the 'local level' even seems to describe the district level, whereas in other parts of the document the district level is separate from the local one. As a result, the *Botswana Climate Change Policy* does not lay out a specific plan for including the *kgotla* in climate change governance.

As a member of the UNFCCC, Botswana must regularly provide updates on its climate change governance, which will then be shared with the international community. This international audience of the report may explain why the institution of the *kgotla*, which is very specific to Botswana's national context, is not mentioned in *Botswana's Third National Communication to the United Nations Convention on Climate Change* (MoENRCT, 2019). However, throughout this report, the idea of vertical integration of climate change adaptation from the national to the local level, which has already been identified as an ambiguous term in this context, is only discussed once. Therefore, the possible localisation of climate change governance in Botswana, while mentioned in national documents, is not represented in documents that are shared internationally.

This communication report (MoENRCT, 2019) does mention the need for further research on and the inclusion of indigenous knowledge systems in climate change matters. However, in the report, the responsibilities are transferred to the Botswana Red Cross Society and the Botswana Climate Change Network, both non-governmental organisations. While the Botswana government sees the need for more information on indigenous knowledge and practices, it shifts the responsibility away from any level of government. The funding and execution of such plans are dependent on external actors and (international) beneficiaries rather than the Botswana government. For example, according to the Botswana Red Cross Society, the organisation receives funding from donors such as the Japanese government, the European Commission of the European Union and the Kuwait Red Crescent Society (Botswana Red Cross Society, 2024). This international funding structure of the Botswana Red Cross Society illustrates the internationalisation of this approach by the government rather than the envisioned localisation of climate action.

To summarize, while some documents at least reference semi specific contexts such as indigenous practices regarding forest management, the majority of governmental output regarding the concept remains vague. There is the constant promise to include the concept of indigenous knowledge, which is not actually defined in any way, in decision making processes and policies. The lack of definition by the government turns the concept into an empty signifier, which is simply used to gain legitimacy in local communities but does not actually work towards equalising different epistemologies.

Real Life Implementation

According to a survey by *Afrobarometer* conducted in 2022, only 29% of Batswana have heard of the phenomenon of climate change (Afrobarometer, 2023). While this number does not give us any direct insights into the role of the *kgotla* in Botswana's climate change governance, it does suggest that the *kgotla*, a forum also used to disseminate information to the communities, is not effectively embedded in Botswana's climate communication strategy. When analysing this survey, it is important to note that the Setswana phrase for climate change, 'maemo a bosa', is also used when talking about the weather, which complicates climate communication.

Moreover, 80% of the people aware of climate change in Botswana believe that ordinary people can act in ways to mitigate the effects of the climate crisis (Afrobarometer, 2023). This belief in small-scale actions leading to large-scale change can be interpreted to mean that the people of Botswana do trust local actors, such as the *kgotla*, to get involved in climate action. Furthermore, elsewhere, this chapter discusses the struggle for political power between the *kgotla* and the national government. In order to remain politically relevant in the contemporary, postcolonial political landscape, it is in the interests of the *kgotla* to play a key role in climate change governance, one of the most defining policy fields for the livelihood of Botswana citizens.

An example of Botswana's local climate change governance to consider is the country's disaster management system. Botswana, with its semi-arid climate, is a country heavily affected by droughts. In 2015–2016, the country suffered the most significant drought in over three decades, which led to only 14.6% of required cereals being produced, 20% livestock deaths, and severe water scarcity (Omari Motsumi et al., 2023). Since then, Botswana has developed a *National*

Drought Plan (Batisani, 2020) with specific funds allocated towards the systematic response to droughts. Even though the Office of the President is generally responsible for disaster management, the operational responsibility is enacted on the village-level through the Village Development Committees (Omari Motsumi et al., 2023), which are elected by the *kgotla*. However, Kulthoum Omari Motsumi, Gina Ziervogel and Mark New (2023), in their study on drought governance in Botswana, found that the officially decentralised structure of governance is not effectively used in the event of disaster. The drought management strategies are complex, and expectations are unclear, leading to overwhelmed local authorities not producing the needed results for effective governance (Omari Motsumi et al., 2023). Furthermore, Omari Motsumi and colleagues (2023) explain that the *kgotla* is not part of the planning process of disaster management strategies and that the consultation meetings are seen as 'a courtesy visit to validate the district reports' (Omari Motsumi et al., 2023: 8). Hence, the authors argue that while the political mandate for local governance is in place, collaborative efforts must be made to '[move] away from tokenistic approaches to local participation' (Omari Motsumi et al., 2023: 11). This analysis of the policy process reveals that the vertical integration Botswana strives for in its *National Adaptation Framework Plan for Botswana* (MoENRCT, 2020) is currently not being implemented in real-life climate change governance.

Another example of localised climate change governance in Botswana is community-based natural resource management. Even though community-based organisations, such as the Village Development Committees, were in charge of implementing projects related to natural resource management, 65% of any revenue created from the projects went towards a national conservation fund, and only 35% of revenue remained at the local level (Shinn, 2016). The government claimed this was due to mismanagement by the community-based organisations, but there are allegations that these claims were exaggerated in order to maintain power and financial control over the projects (Shinn, 2016). The relationship between the government and the *kgotla* that emerged from this has been compared to that of a parent and an older child, where there is no coercion but a strong suggestion to do things a certain way (Shinn, 2016).

The final example of climate change governance relates to natural resource management, specifically in the Okavango Delta. Historically, the local communities there have organised their resource management through the *kgotlas,* guided by shared rules based on long-practiced indigenous knowledge about the management of natural resources (Mogende and Kolawole, 2016). This includes only hunting certain animals at specific times, maintaining a nomadic lifestyle allowing plant-based resources to regrow naturally over time and communicating between different communities to ensure fair access for all (Mogende and Kolawole, 2016). Today, there is national legislation in place regulating the use of natural resources, leading to frustration in local communities as the new system of co-managing is seen as less effective (Mogende and Kolawole, 2016). Furthermore, due to the creation of national parks and conservation areas, some tribal communities lost access to their land and resources, damaging their livelihood (Mogende and Kolawole, 2016). Within the new system, local communities have 'user rights' to natural resources and wildlife in the area, but the national government remains in control over the common resource pool, cutting out local communities from decision-making processes (Mogende and Kolawole, 2016). According to the Wildlife Conservation and National Parks Act of 1992, the Land Boards need to be consulted before making any decisions on land allocation and usage (Chapter 38:01. Wildlife Conservation and National Parks Act, 1992). Since the Land Boards are made up of representatives elected by the *kgotla* and ones appointed by the Ministry of Land Management, Water and Sanitation Services, the interests of the local communities are partly represented in this decision-making process. Yet, it is unclear how freely the Land Boards can oppose such legislation when it is implemented by the national government.

Theoretically, on an institutional level, climate change governance in Botswana, in regional and national plans and policies, lays the groundwork for creating a close partnership between the government and the local communities. Yet, even with the systems in place to decentralise and localise climate action, they are not used in the intended way. Rather than empowering local communities and working with the indigenous leadership, the reality of Botswana's climate change governance retains power within the centralised government.

Conclusion

Botswana is often seen as a governance success story in Southern Africa due to its stable democracy, the country's social welfare system, and the steady reduction of poverty levels in Botswana over previous decades (Worldbank, 2024). Much of this success is attributed to the decentralised nature of the political system in Botswana, which aims to value inputs from local communities and empower their voices on a national scale. In climate change governance, specifically, the government maintains promises of this inclusion of indigenous knowledge and leadership in decision-making in their final *A National Climate Change Strategy for Botswana* (UNDP, 2018). In general, public and official communication stresses the importance of including the *kgotla* and indigenous knowledge in their outputs.

However, in reality, local communities in Botswana are subjected to wilful hermeneutic ignorance when their indigenous knowledge is not included in the epistemic resource pool of the national government. It is unclear if the institutionalised efforts to include indigenous knowledge are simply a performative act towards local communities or if indigenous inputs will actually be valued in climate change-related matters in Botswana.

This question remains especially because when analysing the political system and the local climate change governance in Botswana, there is a stark discrepancy between the institutionalised signalled intentions through reports and policies and the real-life experiences of local communities. The *kgotla* lost much of its governing power after Botswana gained independence and restructured the country in an attempt to decolonise all parts of society. Many of the authority and responsibilities previously in the hand of the *kgosi* and the *kgotla* are now a part of the democratically elected government. The remaining political responsibilities of the *kgotla* are heavily controlled by the government. For example, land ownership was traditionally part of the *kgotla's* domain. Today, it is regulated through the Land Boards or the *dikgosi* themselves, who can be suspended by ministers should they not act accordingly to the government's wishes.

The government recognises the high level of legitimacy the *kgotlas* enjoy in the villages and uses this to increase their legitimacy instead of using the *kgotla* as a direct line to the people. The *Ntlo ya Dikgosi* on the national level represents the indigenous leadership's approval of the government's actions and policies. Yet, the *dikgosi* who are part of the *Ntlo ya Dikgosi* cannot actually influence the decision-making process and are only able to consult. Hence, including the *dikgosi* through the *Ntlo ya Dikgosi* is a performative act by the government. This is also reflected on the local level. The tokenisation of the *kgotla* to legitimise government projects has been well documented above. A minister touring the country before elections without really addressing the concerns of the citizens while getting positive media coverage for his actions is just one example showing the lack of meaningful consideration local communities receive from the government (Obasi and Lekorwe, 2014). More importantly, actual climate change policies, which are designed to include the *kgotla*, are purposively misconstrued to keep community involvement minimal.

It is, however, also important to note that the *kgotla* itself is not an unproblematic leadership system. While it does provide direct access to communities, there are still (invisible) barriers to participation for certain groups of society, namely already marginalised groups such as women or people of low socio-economic status. Moreover, the opinions of the *kgosi*, people close to them, and elderly and male members of the community are disproportionally regarded with higher value simply due to their position in society. Hence, the *kgotla* itself has been facing criticism of being consultative rather than participatory. This also raises the question of whether the *kgotla* should be the local institution responsible for localising climate change governance since the forum is not designed to truly represent the interests of all Botswana citizens but rather upholds patriarchal standards.

Overall, it can be said that local climate change governance in Botswana is not fully incorporating indigenous leadership structures, although the institutional framework exists. Rather than addressing hermeneutic epistemic injustice by including indigenous knowledge in climate change governance, the Botswana government seems to be merely tokenising indigenous communities and the *kgotla* to gain legitimacy. While Botswana's decentralised government can be seen as a good model regarding political structures to ensure localised climate change governance, the real-life implementation of such governance is severely lacking. Moving forward, the Botswana government should make use of the many existing local structures, aim to empower local communities, and tap into the valuable resource of indigenous knowledge available to them. In what capacity the *kgotlas* specifically should be involved in this process remains an open question where the government needs to balance including all voices of society equally into their policy process while also honouring the well-established system of the *kgotla*, which still enjoys high levels of support in rural communities.

References

Afrobarometer. (2023). *Climate Change in Botswana. The People's Perspective*. Afrobarometer. Retrieved September 19, 2024, from https://www.afrobarometer.org/wp-content/uploads/2023/03/Botswana.png

Alexander, K., Carzolio, M., Goodin, D., & Vance, E. (2013). Climate change is likely to worsen the public health threat of diarrheal disease in Botswana. *International Journal of Environmental Research and Public Health*, 10(4), 1202–1230. https://doi.org/10.3390/ijerph10041202

ASEAN Secretariat. (2021). *ASEAN State of Climate Change Report*

Batisani, N. (2020). *National Drought Plan*. UNCCD. Retrieved October 23, 2024, from https://www.unccd.int/sites/default/files/country_profile_documents/National%20Drought%20Plan%20-%20BOTSWANA.pdf

Bond, M. (2010). Localizing climate change: Stepping up local climate action. *Management of Environmental Quality: An International Journal*, 21(2), 214–225. https://doi.org/10.1108/14777831011025553

Botswana Government. (2016). *Botswana Intended Nationally Determined Contribution*. UNFCCC. Retrieved June 25, 2024, from https://unfccc.int/sites/default/files/NDC/2022-06/BOTSWANA.pdf

Botswana Red Cross Society. (2024). *Botswana Red Cross Society—Home*. Retrieved September 17, 2024, from https://www.botswanaredcross.org.bw/

Brown, M. A., Dwivedi, P., Mani, S., Matisoff, D., Mohan, J. E., Mullen, J., Oxman, M., Rodgers, M., Simmons, R., Beasley, B., & Polepeddi, L. (2021). A framework for localizing global climate solutions and their carbon reduction potential. *Proceedings of the National Academy of Sciences*, 118(31). https://doi.org/10.1073/pnas.2100008118

Chieftainship Act, 1987. (1987). Retrieved October 23, 2024, from https://botswanalaws.com/StatutesActpdf/1987Actpdf/CHIEFTAINSHIP%20ACT,%2019%20OF%201987.pdf

Ciaffa, J. A. (2008). Tradition and modernity in postcolonial African philosophy. *Humanitas: Interdisciplinary Journal*, 21(1–2), 121–145

Dube, O. P. (2003). Impact of climate change, vulnerability and adaptation options. Exploring the case for Botswana through Southern Africa. A Review. *Botswana Notes and Records*, 35(1), 147–163

Government of Botswana. (2008). Bogosi Act. Retrieved July 28, 2025 from https://faolex.fao.org/docs/pdf/bot91578.pdf

Government of Botswana. (2018). Tribal Land Act, 2018. Retrieved July 28, 2025 from https://faolex.fao.org/docs/pdf/bot204211.pdf

Helle-Valle, J. (2002). Seen from below: Conceptions of politics and the state in a Botswana Village. *Africa*, 72(2), 179–202. https://doi.org/10.3366/afr.2002.72.2.179

Ifezue, G. O. (2015). Understanding chieftainship in Botswana: The status and powers of chiefs in present day Botswana. *Journal of African Law*, 59(2), 271–293. https://doi.org/10.1017/S0021855315000157

Makgala, C. J., Chebanne, A., Manatsha, B. T., & Sesa, L. L. (2020). The discourse of tribalism in Botswana's 2019 general elections. *Alternation – Interdisciplinary Journal for the Study of the Arts and Humanities in Southern Africa, SP36*.https://doi.org/10.29086/2519-5476/2020/sp36a10

Ministry of Environment, Natural Resources, Conservation and Tourism [MoENRCT]. (2019). *Botswana's third national communication to the United Nations framework convention on climate change*. UNFCCC. Retrieved September 17, 2024, from https://unfccc.int/sites/default/files/resource/BOTSWANA%20THIRD%20NATIONAL%20COMUNICATION%20FINAL%20.pdf

Ministry of Environment, Natural Resources, Conservation and Tourism [MoENRCT]. (2020). *National adaptation plan framework for Botswana*. Government of Botswana. Retrieved September 17, 2024, from https://napglobalnetwork.org/wp-content/uploads/2020/06/napgn-en-2020-nap-framework-for-botswana.pdf

Ministry of Environment, Natural Resources, Conservation and Tourism [MoENRCT]. (2021). *The Botswana climate change policy*. Republic of Botswana. Retrieved June 22, 2024, from https://www.bb.org.bw/common_up/business-botswana/doc_1713877425.pdf

Mogende, E., & Kolawole, O. (2016). Dynamics of local governance in natural resource conservation in the Okavango Delta, Botswana. *Natural Resources Forum*, 40(3), 93–102. https://doi.org/10.1111/1477-8947.12098

Mogomotsi, P. K., Mogomotsi, G. E. J., & Hambira, W. L. (2018). Paris agreement on climate change and Botswana's Vision 2036: An examination of linkages. *Chinese Journal of Population Resources and Environment*, 16(1), 59–66. https://doi.org/10.1080/10042857.2018.1438000

Molebatsi, C., & Morobolo, S. B. (2021). Reading the place and role of endogenous governance structures in modernist physical planning: The case of the *Bogosi* and the *Kgotla* in Botswana. *African Studies*, 80(2), 134–152. https://doi.org/10.1080/00020184.2021.1937057

Monageng, M. (2006). How Britain underdeveloped bechuanaland protectorate. A brief critique of the political economy of colonial Botswana. *Africa Development*, 31(1), 66–88

Mpandeli, S., Naidoo, D., Mabhaudhi, T., Nhemachena, C., Nhamo, L., Liphadzi, S., Hlahla, S., & Modi, A. T. (2018). Climate change adaptation through the water-energy-food nexus in Southern Africa. *International Journal of Environmental Research and Public Health*, 15(10), 2306. https://doi.org/10.3390/ijerph15102306

Mugari, E., Masundire, H., & Bolaane, M. (2020). Adapting to climate change in semi-arid rural areas: A case of the Limpopo Basin part of Botswana. *Sustainability*, 12(20), 8292. https://doi.org/10.3390/su12208292

Mungure, M.-L. D. (2024, October 9). Democracy in Action: The role of civil society in Botswana's failed constitutional amendment bill. *ConstitutionNet*. Retrieved February 26, 2025, from https://constitutionnet.org/news/voices/civil-society-botswanas-failed-constitutional-amendment-bill

National Planning Commission, Office of the President. (2023). *Second transitional national development plan. April 2023–March 2025.* Republic of Botswana. Retrieved August 11, 2024, from https://www.npc.gov.bw/sites/default/files/resources/Second%20Transitional%20National%20Development%20Plan%20-%20FINAL.pdf

Ngwenya, B. N., & Kgathi, D. L. (2011). Traditional public assembly (Kgotla) and natural resources management in Ngamiland, Botswana. In *rural livelihoods, risk and political economy of access to natural resources in the Okavango Delta, Botswana* (pp. 249–262). Nova Science Publishers, Inc.

Obasi, I., & Lekorwe, M. (2014). Citizen engagement in public policy making process in Africa. The case of Botswana. *Public Policy and Administration Research*, 3(4), 1–11

Omari Motsumi, K., Ziervogel, G., & New, M. (2023). Drought governance: A cross-level governance analysis in Botswana. *Climate Risk Management*, 42, 100557. https://doi.org/10.1016/j.crm.2023.100557

Patriot. (2021, January 16). Masisi, cabinet molested Kgotla system -Moatlhodi. *The Patriot*. Retrieved February 26, 2025, from https://thepatriot.co.bw/masisi-cabinet-molested-kgotla-system-moatlhodi/

Phago, K., & Molosi-France, K. (2018). Reconfiguring local governance and community participation in South Africa and Botswana. *Local Economy: The Journal of the Local Economy Policy Unit*, 33(7), 740–756. https://doi.org/10.1177/0269094218809615

Pohlhaus, G. Jr. (2012). Relational knowing and epistemic injustice: Toward a theory of willful hermeneutical ignorance. *Hypatia*, 27(4), 715–735

SADC. (2014). *Protocol on Environmental Management for Sustainable Development*. Retrieved October 23, 2024, from https://static.pmg.org.za/180220Protocol_o.pdf

SADC. (2015a). *Regional Agriculture Policy, the SADC Science, Technology and Innovation Climate Change Response Framework*. SADC. Retrieved October 23,

2024, from https://www.sadc.int/sites/default/files/2021-11/SADC_Climate_Change_Strategy_and_Action_Plan-English.pdf

SADC. (2015b). *Regional Green Economy Strategy for Sustainable Development and Action Plan*. SADC. Retrieved October 23, 2024, from https://www.sadc.int/sites/default/files/2021-11/SADC_Green_Economy_Strategy_and_Action_Plan-English.pdf

SADC. (2015c). *SADC Climate Change Strategy and Action Plan*. SADC. Retrieved October 23, 2024, from https://www.sadc.int/sites/default/files/2021-11/SADC_Climate_Change_Strategy_and_Action_Plan-English.pdf

Seleka. (2018). Integrating indigenous knowledge into rural development in Botswana: A case study. In *Agenda 2063: Culture at the heart of sustainable development* (Vol. 2, pp. 21–35). HRSC Press

Sharma, K. C. (2005). *Role of Traditional Structures in Local Governance for Local Development. The Case of Botswana*. World Bank. Retrieved June 27, 2024, from https://capacity4dev.europa.eu/library/sharma-kc-n-d-role-traditional-structures-local-governance-local-development-case-botswana_en

Shinn, J. E. (2016). Adaptive environmental governance of changing social-ecological systems: Empirical insights from the Okavango Delta, Botswana. *Global Environmental Change*, 40, 50–59. https://doi.org/10.1016/j.gloenvcha.2016.06.011

Statistics Botswana. (2014). *Population and housing census 2011. Analytical report. Statistics*. Botswana. Retrieved August 31, 2024, from https://microdata.statsbots.org.bw/index.php/catalog/15/download/87#page=173

Swemmer, A. M., Bond, W. J., Donaldson, J., Hempson, G. P., Malherbe, J., & Smit, I. P. J. (2018). The ecology of drought – a workshop report. *South African Journal of Science*, 114(9/10). https://doi.org/10.17159/sajs.2018/5098

Tlou, J. S. (2021). Ubuntu as Reflected in the Kgotla System of the Government in Botswana. In J. Mukuni, & J. Tlou (Eds.), *Advances in Religious and Cultural Studies* (pp. 1–8). IGI Global. https://doi.org/10.4018/978-1-7998-7947-3.ch006

UNDP. (2018). *A national climate change strategy for Botswana*. UNDP. Retrieved June 22, 2024, from https://faolex.fao.org/docs/pdf/bot208159.pdf

Wildlife Conservation and National Parks Act, Chapter 38:01 (1992). Retrieved August 11, 2024, from https://faolex.fao.org/docs/pdf/bot4728.pdf

World Bank. (2024). *Botswana Country Profile*. World Development Indicators Database. Retrieved August 31, 2024, from https://databank.worldbank.org/views/reports/reportwidget.aspx?Report_Name=CountryProfile&Id=b450fd57&tbar=y&dd=y&inf=n&zm=n&country=BWA&_gl=1*1n0cuju*_gcl_au*NTk5NTA3MzI4LjE3MTgyNjY4NDA

Decolonial Reflexivity and Climate Change Scholarship in Botswana

7

Gabriel Faimau

Abstract

The discourse of decolonisation and decoloniality has generally focused on examining the hegemony of the Western knowledge systems, unmasking the Western epistemological tradition and reframing the quest for knowledge production in the context of a pluriversality of epistemic traditions and cultures. In doing so, the decolonial turn in the global entanglements of knowledge has successfully diagnosed the embedded epistemic violence in the Western epistemic tradition, which leads to what has been characterised as 'epistemicide'. However, the decolonial enterprise seems to rely on the basic assumption that a new knowledge system can easily be developed in place of a dominant system of knowledge production. In this chapter, I open a new path within the decolonial turn by adopting the idea of decolonial reflexivity that banks on a conversational and dialogic approach to the co-production of knowledge. To illustrate the propositions of decolonial reflexivity empirically in the context of knowledge co-production, I turn to climate change research and representation in Botswana to demonstrate the ways in which decolonial reflexivity could be applied in climate change scholarship and discourse.

G. Faimau (✉)
Department of Sociology, University of Botswana, Gaborone, Botswana;
Unit for Reformational Theology and the Development of the South African Society, North-West University, Potchefstroom, South Africa
e-mail: FaimauG@ub.ac.bw

A. C. Terblanché Greeff et al. (eds.), *Climate Change and the Postcolonial*, Kulturelle Figurationen: Artefakte, Praktiken, Fiktionen,
https://doi.org/10.1007/978-3-658-49328-8_7

Keywords

Decolonisation · Decoloniality · Decolonial reflexivity · Climate change · Knowledge co-production · Botswana

Introduction

In the past few decades, decolonial scholarship has focused on the discourse of decoloniality to examine the dark side of epistemology that has dominated the epistemic experiences of the Global South. If coloniality creates a power structure and an epistemological design to maintain the hierarchical imbalances and imperialist model of knowledge that is based on the binary thinking of primitive vs. modern/civilised, decoloniality unpacks the alienation that has disempowered indigenous knowledges and pushed them away to the margins of society (Ndlovu-Gatsheni, 2013; 2015; 2020). Decoloniality, therefore, challenges epistemic violence and 'epistemicide' that has been produced through the colonial power and knowledge structure (Santos, 2014). While decoloniality offers a way of examining the hegemony of the Western knowledge systems, unmasking the Western epistemological tradition, and exposing the ills of coloniality and its effects (Matolino, 2020), the discourse seems to rely on the basic assumption that a new knowledge system can be easily developed in place of a dominant system of knowledge production.

This chapter offers a new path within the decolonial turn by adopting the idea of decolonial reflexivity that banks on a conversational and dialogic approach to the co-production of knowledge. Decolonial reflexivity is premised on the idea that dialogue leads to understanding and creates new possibilities. Following this argument, in this chapter I attempt to show how decolonial reflexivity allows for knowledge co-production in the context of the Global South by reflecting on the climate change research and discourse in Botswana. In so doing, I argue that climate change scholarship requires decolonial reflexivity as it calls for constant dialogue with different sources of knowledge, accommodation of experiential knowledge and continuous reflection on lived realities and lived experiences.

The first two sections show an attempt to unpack decoloniality as a school of thought and outline the propositions of decolonial reflexivity. This is followed by a reflection on decolonial reflexivity in conjunction with the notion of co-production. The last two sections offer an analysis of the current state of climate change scholarship in Botswana and the adoption of decolonial reflexivity in the context of climate change research.

Decoloniality and Its Epistemological Construction

At the heart of the decoloniality project is an attempt to develop an analytical approach to questioning the notion that Western epistemology and modes of thinking are universal. Taking a closer look at this notion is crucial as it is believed that coloniality exists as 'an embedded logic that enforces control, domination, and exploitation disguised in the language of salvation, progress, modernization, and being good for everyone' (Mignolo, 2005: 6). The construction of Western epistemology seems to follow the Cartesian worldview that banks on a dichotomic divide between progress and backwardness, the centre and the periphery, the subject and the object, the controller and the controlled, the coloniser and the colonised. This epistemology is characterised by two opposing concepts, namely modernity, which promises progress and development and colonialism, which reflects racism and oppression (Ndlovu-Gatsheni, 2015; Domptail and De Santo, 2023). This epistemological construction seems to offer a basis for understanding the differences and pluralities, particularly within Western thought. However, the line of thinking embedded in it clearly has a broad consequence whereby the 'other' is represented as a 'fixed' entity from an opposing site. Therefore, Homi K. Bhabha (2009: 94) suggested that 'an important feature of colonial discourse is its dependence on the concept of "fixity" in the ideological construction of otherness'.

Different phrases such as 'epistemic disobedience' (Mignolo, 2011) and 'epistemic reconstruction' (Quijano, 2007: 176) have been used by decolonial scholars to characterise the decoloniality project and how it problematises the hegemonic Eurocentric modernity, knowledge system and systemic view of power. Decoloniality, therefore, attempts to redefine and redesign culture, human and intercultural relationships, epistemology and knowledge production (Anugwom, 2023). It challenges the existence of coloniality in various socio-cultural and socio-political institutions and unmasks the embedded hegemonic logic that is designed to dehumanise those who do not belong to the Western geopolitics of knowledge (Mignolo, 2005). In other words, decoloniality provides a liberating framework that creates an opportunity to articulate the epistemologies of the Global South and de-hegemonise the Western knowledge system and epistemological domination (Ndlovu-Gatsheni, 2015; Ndlovu-Gatsheni, 2020; Anugwom, 2023).

According to Sabelo J. Ndlovu-Gatsheni (2013; 2015), one of the leading African decolonial voices, decoloniality builds on three key concepts. The first concept relates to the coloniality of power. Through the coloniality of power, global politics is constructed through a hegemonic lens of a power structure that divides the world into the world of those in charge of the global political arena benefitting from modernity, and another world of incomplete beings or none-beings governed by and through imperialism and colonialism. The second is premised on the coloniality of knowledge with a focus on the politics of knowledge production. Within this context, decoloniality is interested in examining how knowledge production has been predominantly Western-centric and its long-term impact on the Global South. As such, decoloniality questions the relevance of knowledge generation and the extent to which it empowers local communities within which knowledge is produced. The third one is based on the coloniality of being. Here, decoloniality unmasks how coloniality produces the 'coloniser vs. colonised' binary in which the first is considered as the subject, the human and the powerful, while the latter is perceived as the object, the inhuman and the powerless. As such, the humanity of the latter is questioned, undermined or even dismissed. Decoloniality resists objectification, restores a sense of humanity and advocates for a sovereign subjectivity. In this context, decoloniality underscores the agency of those who are dehumanised through a colonial mindset.

Although decoloniality is successful in diagnosing the problem of colonisation and coloniality, the pragmatic understanding of decoloniality requires clarification. There are at least two broad viewpoints relating to decoloniality. The first can be characterised as an ideological view of decoloniality. This view is based on a position that colonial domination does not only shape and influence our ideas of race, class, gender and culture, but also involves 'epistemicide, the murder of knowledge' (Santos 2014: 92). According to Santos (2014), epistemicide implies the death of knowledge as a result of unequal encounter and exchanges among cultures, and the death of knowledge implies the death of socio-cultural groups who possessed it. Therefore, the task of decoloniality is about dismantling Western epistemology, Eurocentric understanding of the world and the monopoly of knowledge system interpretation, or, as suggested by Maldonado-Torres (2006: 117), decoloniality refers to 'the dismantling of relations of power and conception of knowledge that foment the reproduction of racial, gender, and geo-political hierarchies that came into being or found new and more powerful forms of expression in the modern/colonial world'. While maintaining a resistance of Western epistemological domination and acknowledging the problem of colonial history, the second viewpoint favours an epistemic reconstruction that involves 'the reconstruction of a critical imaginary and the rebuilding of a new horizon of future' (Quijano, 2002: 84). In a reconstructive context, decoloniality involves 'processes of "rethinking thinking", "unthinking", and "learning to unlearn in order to relearn"' (Ndlovu-Gatsheni, 2019: 490). The reconstruction process involves acknowledgment that colonial history is not the only history that shapes the knowledge system; other histories or historical processes also play a critical role in the production of knowledge (Landman and Makakavhule, 2021). The idea of reconstruction, however, is often criticised as a model that does not provide a real critical option. According to Mignolo (2014), reconstruction is a manifestation of the imperial expansion of Western thinking as it only leaves two options: being assimilated to the variants of Western modernity, or reconstructing categories of thought from one's local and particular tradition.

Decoloniality is appealing because it poses critical questions and offers a new way of thinking and interrogating the common assumptions of power structure, knowledge production and our view of being. The decolonial approach also values a public recognition of how knowledge and methods of knowledge production are developed through colonial interpretation of knowledge systems. However, decoloniality seems to assume that the dominant group in the global political constellation can simply be persuaded through intellectual argumentation and moral appeal (Parekh, 2001). It also assumes that a new conception of knowledge can be easily developed in place of the dominant system of knowledge generation. In addition, decoloniality seems to overlook alternative strategies that could benefit the Global South, such as capitalising on the indigenous knowledge systems, strengthening regional integration, and enhancing South-South cooperation (Matolino, 2020). As such, any analysis employing a decolonial approach cannot escape two fundamental questions: first, what alternative epistemologies does the decolonial project offer? Second, whose interests are served by a decolonial analytical approach? It is not the intention of this chapter to respond to these questions. Nevertheless, in the sections below I intend to offer another way of understanding decoloniality that is accommodating of a multicultural context and local epistemologies.

Towards a Decolonial Reflexivity

Decoloniality underscores epistemic and political liberation because one of its key features is the acknowledgment of the existence of different epistemologies or nondominant knowledges. It is therefore often suggested that 'epistemological decolonization requires taking nondominant knowledges and their epistemes seriously to open up the possibility of interrogating and dismantling the hegemony of the Western knowledge tradition' (Botha et al., 2021: 51). Decolonial scholars believe that cultures are spaces and loci of thinking. As such, the task of thinking, theorising or reasoning is universal in nature and not exclusive to a particular culture. Among decolonial scholars, this view is underlined through a reflection on three social realities: the ecology of knowledges, pluriversality, and interculturality. Introduced by Santos (2014), the phrase 'ecology of knowledges' refers to 'the recognition of the different ways of knowing by which people across the human globe provide meaning to their existence and understanding of the world' (Ndlovu-Gatsheni, 2019: 204). Ecology of knowledges thus confronts the notion and the logic of monocultural thinking in the production of scientific knowledge. Pluriversality signifies a vision of co-existence and co-operation of many worlds in a pluriverse. As such, 'the argument for decoloniality must at the same time work to wrestle decoloniality from the temptations of totalitarian totality. Decoloniality promotes pluriversality as a universal option—which means that what "should be" universal is in fact pluriversal, and not a single totality' (Mignolo and Walsh, 2018: 147). The third idea, interculturality, is not simply about dialogue or encounter between cultures. Interculturality is about building different societies with new structural ordering, socio-political and cultural transformation (Mignolo and Walsh, 2018).

Although decoloniality relies on the proposition that reason manifests itself in different (or all) cultures, a practical operationalisation of this idea has generally been overlooked. Therefore, I suggest 'decolonial reflexivity' as a strategic positioning in understanding decoloniality, particularly in the context of the relationships between and among cultures. The term decolonial reflexivity mirrors the notion of postcolonial reflexivity introduced by Nothias (2020: 246) to characterise 'an ability to recognize the negative contribution of the news industry to representational Othering'. Decolonial reflexivity refers to acknowledging the colonial epistemology that dominates the interpretation of knowledge systems and produces a structure of authority that governs knowledge production. From this point of departure, decolonial reflexivity views epistemological design and knowledge production as a conversational arena for mutual exchange among various epistemologies that reflect the particularities of different societies and cultures. This is to say that decolonial reflexivity favours a conversational or relational tone of decoloniality as an epistemic movement.

The decolonial reflexivity proposal reflects on the recent trends of having dialogue at the centre of decolonial discourses. Boaventura de Sousa Santos (2014), for example, calls for an epistemic diversity that favors intercultural translation. For Santos, intercultural translation does not rely on the binary approach in comparing epistemologies as it questions the dichotomies of epistemologies and the unequal abstract status of knowledge. Instead, it identifies the differences and similarities and develops 'new hybrid forms of cultural understanding and intercommunication that may be useful in favoring interactions and strengthening alliances among social movements fighting, in different cultural contexts, against capitalism, colonialism, and patriarchy and for social justice, human dignity, or human decency' (Santos, 2014: 212). In other words, intercultural translation enables a critical conversation among cultures and calls for respectful dialogue across epistemologies based on a shared commitment to solidarity and global cooperation (Guilherme, 2021). At a practical level, critical conversation involves reflection on and interrogation of an issue from multiple perspectives (Huntsley and Brentnall, 2021) and respectful dialogue affirms the moral significance of human beings as social beings (Taylor, 1994). Similarly, in the context of sociology, Ali Meghji (2021) advocates for a decolonial spirit that rejects the hierarchical view of knowledge systems in favour of collaboration and conversation. While acknowledging the damage that colonial power has done, this viewpoint embraces the position that decolonising is not about diminishing a certain mode of knowledge production or epistemological approach; it is rather about broadening the field of conversation and creating intercultural relationships that allow for cultural encounters and knowledge exchange. Or, as argued by Meghji (2021: 97), decolonising 'does not necessitate a dismantling of "Northern theory", but rather involves a horizontal, reflective, conversational approach'. A similar argument was proposed by Sari Hanafi during the presidential address titled 'Toward a Dialogical Sociology', delivered at the XX International Sociological Association's (ISA) World Congress of Sociology. In this address, Hanafi argued that dialogical sociology offerred 'pluralistic conceptions of the good, balancing individual, and collective rights' (2023: 19). Decoloniality therefore embraces what Jonathan O. Chimakonam calls conversational thinking as a tool for educated and enlightened engagement that aims at correcting and strengthening institutions, transforming the public sphere and creating new ideas and thoughts (Chimakonam, 2017; Egbai and Chimakonam, 2019). One could broadly argue that decolonial reflexivity favours a networked relational knowledge generation model that prioritises the role and importance of relationships among actors, artifacts and spaces in the knowledge generation efforts on the one hand, and creates and nurtures networking and close connections among individuals and communities that make up knowledge-producing communities on the other (Botha et al., 2021).

At a practical level, however, how do dialogue and conversation take place in the context of decolonial reflexivity? The dialogic nature of decolonial reflexivity is centred around knowledge and understanding of differences. Or, as argued by Parekh (2001: 128), human beings 'are similar enough to be intelligible and make a dialogue possible, and different enough to be puzzling and make a dialogue necessary'. Knowledge of the differences, however, does not always require a complete understanding of those differences. This brings us to an argument made by Zali Gurevitch that 'the ability to not understand, rather than the ability to understand the other, is posited as crucial to the dialogic process' (Gurevitch, 1989: 161). According to Gurevitch, generally, our view of understanding the differences is limited to the move from an 'inability to understand' to an 'ability to understand'. The problem is that often when we think we have the 'ability to understand,' we are actually in a condition that can be characterised as an 'inability to not understand.' This condition happens when understanding is used as a cover-up for our inability to understand others and our relationship with others. In this context, the presence of an 'inability to not understand' creates another corner that we may call the 'ability to not understand.' Interestingly, what is required under the 'ability to not understand' is not a complete explanation of the other but rather the ability to familiarise oneself with the unfamiliar terrain or the strangeness of the Other (Faimau, 2013: 93–95).

How do we understand the idea of familiarising oneself with the unfamiliar terrain of the Other? One of the concepts that can help us to respond to this question is the concept of social learning. Social learning refers to a learning process that focuses on the exchanging of knowledge and experience through interaction among human actors (von Schonfeld et al., 2019). This is to say that while acknowledging the differences, human actors have the creative capacity to relate to one another and find common ground that makes dialogue possible. This capacity often involves what philosopher Gabriel Marcel calls 'recollection' or the process of renewing and transforming oneself to 'being-in–a-situation', a situation of being unfamiliar with the other (Sweetman, 2008). As far as social learning is concerned, the interaction (in the context of being-in–a-situation) allows for co-creation as a result of the interplay between an individual and other actors. This means co-creation redefines the power relationships that exist within the interaction process. Therefore, social learning can be considered a communicative strategy for crossing boundaries and barriers in any setting including unfamiliar settings (von Schonfeld et al., 2019).

Decolonial Reflexivity and the Notion of Co-Production

The dialogic approach of decoloniality or decolonial reflexivity favours epistemic justice by laying stress on the idea of co-production of knowledge or epistemological co-production. Co-production is an epistemic praxis that recognises the inter-relationship between knowledge production and the societal context with a strong commitment to holding together knowledge and action (Perry, 2022).

This means that through co-production, knowledge informs actions or practices and actions produce or create knowledge. As such, co-production is often conceptualised as 'a spectrum of practices between macro-knowledge systems and micro-knowledge systems aimed at increased knowledge utilization' (Bandola-Gill et al., 2022: 277). Co-production, therefore, assumes a reciprocal process of knowledge exchange between diverse stakeholders (Durose et al., 2022) as it combines scientific knowledge and local knowledge generated through knowledge sharing among stakeholders (Arifah et al., 2023). In short, knowledge co-production assumes a process of collaboration involving various stakeholders and shareholders that bring together different sources of knowledge and information to generate new knowledge and strategic systems and address a commonly defined problem.

At a practical level, however, I suggest that co-production assumes two related ideas: positionality and intentionality. Positionality refers to the understanding of the socio-political and cultural context that shapes an individual and his/her identity. In other words, the concept invites a cognizance of one's own history and identity that contribute to one's knowledge, perspective and social practices. This basic understanding is important as our epistemological position is often influenced by our own positionality (Trisos et al., 2021; Hill et al., 2023). Intentionality refers to the capacity, mental state and consciousness that direct one's understanding and acceptance of a reality in which diversity, pluriversality and multiple knowledge systems exist. Intentionality assumes that agency alone is not enough. As such, intentionality is required to make it possible for the intended meaning to be produced in co-production of knowledge. In this context, intentionality is relational because its main function is to mediate between different structures and one's agency directed toward knowledge co-production (Li, 2023). In the context of decolonial reflexivity, intentionality initiates a process of unlearning and relearning through which different epistemologies are welcomed and come into play in the co-production process (Hill et al., 2023).

Overview of Climate Change Research in Botswana

As indicated in the introductory section, this chapter reflects on the idea of decolonial reflexivity and its application in the context of the co-production of climate change narratives in Botswana. To contextualise the reflection that will be developed in the final section, this one gives a broad overview of the state of climate change research in Botswana without the intention of developing a detailed bibliometric analysis. This broad overview relies on a literature topic search using the Web of Science Core Collection. As of February 2024, the Web of Science Core Collections shows 703 results on climate change in Botswana, of which 641 are research articles. While there are many articles published in science journals, in this chapter I consider only the climate change research in Botswana, from the perspective of social sciences and the humanities, and I focus on 110 articles indexed in the Social Sciences Citation Index (SSCI) and Arts and Humanities Citation Index (AandHCI). At the time of publication, 89 (co-)authors of these articles were Botswana-based researchers from the University of Botswana, Botswana University of Agriculture and Natural Resources (BUAN) and Botswana International University of Science and Technology (BIUST).

Although the articles address many different research areas of climate change in Botswana, the majority of them focus on the impact of climate change on the following sectors: (1) agriculture, farming and food security; (2) wildlife and tourism; (3) water supply and resource management; (4) public health; and (5) ecosystem services. To demonstrate how researchers have discussed the issue of climate change and its impact on various sectors in Botswana, I will provide examples from the first two sectors listed above. In the agricultural, farming and food security sectors, the authors generally focus on the risks climate change poses to agricultural productivity and food security. Mogomotsi, Skelemani and Mogomotsi (2020), for example, raise the issue of climate change and variability in relation to the sustainability of agricultural and food production by examining factors that shape farmers' adaptation to climate change and the climate change adaptive capacity of farming households in Ngamiland East. Wildlife and tourism are important sectors that contribute substantially to Botswana's economy. There is no doubt that these sectors have been badly affected by the frequency of extreme events such as heat waves and drought in Botswana. While acknowledging the variability of Botswana's key climate elements, researchers also argue for proactive thinking of economic diversification and tourism policies that are responsive to the (future) uncertainty of climate change impacts (Saarinen et al., 2012; Hambira et al., 2020; Saarinen et al., 2020).

Apart from these areas, many of the articles reflect on thematic areas such as the history of climate change in Botswana, knowledge and perception of climate change as well as the attitudes towards it, climate change policy and adaptation strategies, the role of the indigenous knowledge systems, and gender and the effects of climate change. In the area of indigenous knowledge systems, researchers value the importance of these systems in the sustenance of the local communities and the contribution of local knowledge to environmental and climate change knowledge. As such, they suggest a critical consideration of indigenous knowledge systems in climate change adaptation and disaster risk reduction strategies and policies (Dintwa et al., 2022). Similarly, articles that focus on gender and climate change call for scholarly research on the gender dynamics around the climate adaptation priorities and practices at the household level, and for the development of a comprehensive climate adaptation strategy that is informed by local memories and historical knowledge of environmental change adaptation (Ketlhoilwe, 2013; Ngwenya et al., 2017; Makwatse et al., 2022).

Decolonial Reflexivity and Climate Change Research

In the previous section, I provide a brief overview of the climate change research in Botswana from the perspective of social sciences and the humanities. Climate change is often perceived as an abstract concept. The conceptual abstractness of climate change often diminishes the urgency felt by social actors such as policy makers, scientists and researchers, civil society and advocacy groups, and general public, leading to a lack of meaningful engagement in addressing this critical threat. In 2023, Jörn Ahrens and Ernst Halbmayer published an interesting edited volume titled *Climate Change Epistemologies in Southern Africa: Social and Cultural Dimensions* (Ahrens and Halbmayer, 2023). Introducing the book, they argue that understanding the challenge of climate change and its socio-cultural and environmental impact requires a critical approach to the issue as both a universal and a heterogeneous phenomenon. As such, any analysis of climate change and its broad impact should consider various cultural peculiarities relating to the climate change issue. This assertion highlights the complexity of climate change discourses on the one hand, and urges for cultural sensitivity in the production of climate change knowledge and epistemologies on the other.

In the final section of this chapter, while considering the abstractness of climate change, I would like to reflect on how decolonial reflexivity and the epistemology of co-production are at work as far as climate change research in Botswana is concerned. As a country, Botswana is highly dependent on rain-fed agricultural and natural resources. Therefore, climate change and variability pose a serious threat which is acknowledged in various official documents published by the Botswana government. Botswana's Vision 2036, for example, states that 'global warming and climate change are unequivocal and could dampen a country's desired economic growth and development. As a nation, we recognize the possibility that disasters such as fire, floods and drought could worsen with increased incidents of climatic variability. We therefore take a strong stance to include climate change vulnerability assessments, adaptation and mitigation into our development planning' (The Vision 2036 Presidential Task Team, 2016: 24).

This statement sets the tone for Botswana's commitment to climate change adaptation and mitigation and how the climate change threat is included in the country's development plan. As such, climate change has become an important area of research in the context of Botswana.

How does decolonial reflexivity or dialogic decoloniality apply in the presentation of climate change narratives from the context of Botswana? How are climate change knowledge and discourses produced in Botswana? In what ways have climate change narratives been circulated in Botswana's public sphere? How have such narratives acquired a place in the scholarly research on climate change in Botswana? These questions are relevant in terms of unpacking how climate change narratives have been accommodated in the climate change scholarship and circulated in the public sphere. My overview of the current state of climate change research and public discourse in Botswana reveals four essential themes through which decolonial reflexivity may have been (and can be) applied in the context of climate change research and discourse, namely the provision of space for local or indigenous knowledge of climate change, the centrality of experiential knowledge, a reflection on local realities and the visualisation of climate change.

The phrase 'indigenous or local knowledge' has generally been used to describe 'the understandings, skills and philosophies developed by societies with long histories of interaction with their natural surroundings' (UNESCO). Although it has been understood in different ways, indigenous knowledge can be defined as 'the cumulative body of strategies, practices, techniques, tools, intellectual resources, explanations, beliefs, and values accumulated over time in a particular locality without the interference and impositions of external hegemonic forces' (Emeagwali, 2013: 31). Indigenous knowledge reflects the basic principles of indigenous worldviews, traditions and beliefs. As such, it is often characteristically 'local in scale, transmitted orally, collectively owned, holistic in perspective, and adaptive in nature' (Mistry et al., 2020: 211).

Many studies on climate change in Botswana have provided a space for accommodating local knowledge, particularly in the context of climate change adaptation strategies. Mogomotsi, Skelemani and Mogomotsi (2020) published a study on the climate change adaptation strategies of small-scale farmers in Ngamiland East. While developing a quantitative analysis related to the impact of climate change on subsistence agriculture, the authors collected local knowledge of climate change adaptation strategies. Interestingly, their research participants continue to emphasise a belief that traditional knowledge and practices of climate change adaptation strategies have come to be considered as a generational legacy. One of their participants, for example, has this to say: 'The lessons and skills I obtained from my great-grandparents are helping me to feed my family.

I know how to predict seasonal weather conditions because of those skills. I know when to opt for ploughing small grains as opposed to watermelon. I have taught my children. The lessons are a generational legacy' (Mogomotsi et al., 2020: 453). They conclude that the choice of adaptation strategies of a significant proportion of their respondents is largely influenced by indigenous knowledge, particularly when it comes to seasonal forecasting and identification of the beginning of ploughing seasons. Another study focusing on climate change adaptation in semi-arid rural areas by Mugari, Masundire and Bolaane (2000) found the same trend of relying on traditional knowledge and documented various indigenous and local practices of climate change adaptation. However, the researchers noted that the influence of traditional knowledge on adopting climate change adaptation strategies requires validation and synchronisation with scientific information to enhance the timely adoption of climate change adaptation measures.

The above examples show that research on climate change in Botswana generally acknowledges the importance of indigenous knowledge as a strategy in addressing climate change. One of the requirements of reflexive decoloniality is the co-creation of knowledge. Co-creation of knowledge is about bringing a plurality of knowledge sources to address the climate change issue and problem (Hill et al., 2023). Indigenous knowledge or what is characterised as generational knowledge, is certainly one of the knowledge sources. Experiential knowledge is an important theme discussed in Botswana's climate change scholarship related to decolonial reflexivity. Climate change and its impact are often communicated through knowledge sharing based on lived experiences and lived realities. Knowledge by experience or experiential knowledge is valued because it consists of practical knowledge or know-how (Dings, 2023). In 2017, Akinyemi (2017) published a research article assessing the perspectives of smallholder farmers in Eastern Botswana on climate change and its variability. This research documented the farmers' experiential knowledge about changes in climatic conditions and how the knowledge is used to adapt to climate change and its variability. For example, to reduce the risk of crop failure and minimise the impact of climate change on crop production, farmers adopt a strategy of intercropping or planting different crops simultaneously as their experience taught them that the climate change impacts on crops vary and some crops are perceived as drought-resistant. The intercropping strategy is therefore used to give them alternatives as they face the risk of crop failure. Or in the words of one of the farmers quoted by the researcher: 'some crops will survive and some will not. If one crop fails, at least others are harvested and there will be something to eat' (Akinyemi 2017:359). What is interesting is that experiential knowledge of local actors is essentially relational and therefore dialogic because it covers the 'knowledge *about, with* and *for* their environment and the realms beyond them' (Botha et al., 2021: 60). This experiential knowledge is often shared among the farmers as many believe that climate change is a collective problem and it must be faced collectively. In addition, the sharing of experiential knowledge contributes to the local social cohesion based on the belief that 'we are a family and we look out for each other. We care about each other's yield and produce. That's why we share ideas, news and tips' (Mogomotsi et al., 2020: 456).

Decolonial reflexivity involves dialogue with and reflection on local realities because local realities are lived realities. Lived realities are normally told and shared. Therefore, storytelling is often viewed as a decolonial practice and decolonising method (Samuel and Ortiz, 2021). Scholarly representation of local realities is another theme in Botswana's climate change scholarship and discourse. In a recent study that I conducted with my colleagues (Faimau et al., 2023), we established the astonishing fact that although it is a pressing issue and its urgency has been underlined in various official documents in Botswana, climate change discourse has received limited attention and remains relatively absent in the print media reporting. However, from the scarce print media coverage of climate change, we observed a unique aspect of Botswanan media reporting. While print media reports on climate change in other parts of Africa are often disconnected from African realities (Kay and Gaymard, 2020), Botswanan print media coverage of climate change frequently reflects the local realities and challenges, such as lack of rain, severe drought, heat wave, food security, agricultural and food production, nutrition and public health. Rather than using specialised terminologies to explain the scientific nature of climate change, Botswanan print media opt for everyday language to create stories about the extent to which climate change has impacted people's livelihoods, particularly those of local farmers. Similar trends are evident scholarly research, as numerous studies on climate change in Botswana focus on local realities of key sectors that are mostly affected by climate change impacts, such as agriculture and toories. As such, studies on climate adaptation strategies in Botswana often involve key stakeholders such as farmers, rural and urban communities, and policymakers.

While print media coverage on climate change has been limited, other media platforms in Botswana, such as radio and television, have played a significant role in informing the public and encouraging discourse on climate change through many of their programmes (Akinyemi, 2017; Ngwenya et al., 2017). Therefore, it is fair to say that in Botswana people also learn about climate change through the media. Lately, information about climate change has also been communicated through various social media and online media platforms. However, the phenomenon remains an abstract scientific idea for many (Akinyemi, 2017; Ahrens and Halbmayer, 2023; Faimau et al., 2023). Visual representation of climate change is a communication strategy which gives a familiar face to an unfamiliar phenomenon. The decolonial scholarship emphasises the interconnectedness of human beings and ecosystems (Gothe and De Santolo, 2022; Topa, 2022). Or, as Smith argues, (1999: 148), interconnectedness 'positions individuals in sets of relationships with other people and with the environment … To be connected is to be whole'. Therefore, many indigenous creation stories normally link people to the land, water, places, animals, plants, hills and mountains. Visualisation of climate change not only creates a story about climate change, it also tells a story about the interconnectedness of human beings and the environment. The media, especially the printed media, adopt this strategy and represents climate change through the visualisation of local realities that members of the public are familiar with, such as dry dams, drought, dead animals due to extreme heat, and crop failure. Or, as indicated in our recent work, 'While explaining climate change as a scientific and abstract concept requires substantial knowledge and creative communication skills, a picture of a dry Gaborone dam and an image of a dead animal and helpless animals on dry land capture the realities and present the effects of climate change in a powerful way. Here, the unknown climate change becomes known through images that capture and depict the familiar realities of Botswana' (Faimau et al., 2023: 161). Visualisation of local realities not only gives evidence of the climate crisis but also shows that the crisis is real and its impact cannot be ignored. In addition, visualisation is appealing because it connects and reconnects people with their environment and the ecosystems they are familiar with. It reminds them of their constant dialogue with nature and the environment. Visualisation of climate change therefore reinforces the idea of dialogic decoloniality.

Conclusion

The current quest for decolonisation is primarily concerned with the dismantling of power hierarchies and Western epistemological and intellectual hegemony that decentre other or non-Western ways of knowing. The decolonisation and decoloniality scholarship has been successful insofar as it has identified and diagnosed the problem of Western power structures and hegemonic epistemology. What is still lacking is a clear proposal on how to bring indigenous voices, knowledges or epistemologies to the centre of scholarship and research. This chapter presents the idea of a decolonial reflexivity that underlines a conversational and dialogic approach to knowledge production. I argue that while acknowledging the colonial epistemology and its impact on the scholarship and research, decolonial reflexivity considers epistemological design and knowledge production as a conversational arena for mutual exchanges among 'other' epistemologies that reflect the particularities of different societies and cultures. Using the propositions of this decolonial turn, the chapter further explores how decolonial reflexivity is applied in the context of climate change scholarship in Botswana. A broad overview of scholarly research in Botswana shows that climate change research has generally focused on key sectors such as agriculture, farming and food security, wildlife and tourism, water supply and resource management, public health, and ecosystem services. The analysis of the extent to which decolonial reflexivity may have been applied in climate change research and public discourse in Botswana reveals key thematic areas such as the provision of space for local knowledge of climate change, an emphasis on the centrality of experiential knowledge, a reflection on local realities and the visualisation of climate change.

The present chapter is limited to the exploration of decolonial reflexivity and shows how it may have provided a lens for climate change research and discourse in Botswana. Applying and promoting decolonial reflexivity in research or knowledge generation would therefore imply a creative and ethical endeavour to build inclusive research cultures that acknowledge the plurality of knowledge sources and develop an integrated system of co-production to address a well-identified and well-defined common problem.

References

Ahrens, J., & Halbmayer, E. (2023). *Climate change epistemologies in Southern Africa: Social and cultural dimensions*. London: Routledge

Akinyemi, F. (2017). Climate change and variability in Semiarid Palapye, Eastern Botswana: An assessment from smallholder farmers' perspective. *Weather, climate, and society*, 9(3), 349–365. https://doi.org/10.1175/WCAS-D-16-0040.1

Anugwom, E. (2023). Beyond the debate: decolonisation, decoloniality and the reframing of development in Africa. *Politikon: South African Journal of Political Studies*, 50(4), 419–437. https://doi.org/10.1080/02589346.2023.2280388

Arifah, Salman, D., Yassi, A., & Demmallino, E. B. (2023). Knowledge flow analysis of knowledge co-production-based climate change adaptation for lowland rice farmers in Bulukumba Regency, Indonesia. *Regional Sustainability*, 4(2), 194–202. https://doi.org/10.1016/j.regsus.2023.05.005

Bandola-Gill, J., Arthur, M., & Leng, R. (2022). What is co-production? Conceptualising and understanding co-production of knowledge and policy across different theoretical perspectives. *Evidence and Policy*, 19(2), 275–298. https://doi.org/10.1332/174426421X16420955772641

Bhabha, H. K. (2009). *The location of culture*. London and New York: Routledge

Botha, L., Griffiths, D., & Prozesky, M. (2021). Epistemological decolonization through a relational knowledge-making model. *Africa Today*, 67(4), 51–72

Chimakonam, J. (2017). Conversationalism as an emerging method of thinking in and beyond African philosophy. *Acta Academica: Critical views on society, culture and politics*, 49(2), 11–33. https://doi.org/10.18820/24150479/aa49i2.1

Dings, R. (2023). Experiential knowledge: From philosophical debate to health care practice? *Journal of Evaluation in Clinical Practice*, 29, 1119–1126. https://doi.org/10.1111/jep.13845

Dintwa, K., Letamo, G., & Navaneetham, K. (2022). Vulnerability perception, quality of life, and indigenous knowledge: A qualitative study of the population of Ngamiland West District, Botswana. *International Journal of Disaster Risk Reduction*, 70(102788). https://doi.org/10.1016/j.ijdrr.2022.102788

Domptail, S., & De Santo, M. (2023). Beyond Western dichotomies of power: Life-centered development, reciprocity and co-creation within nature. In M. De Santo, & S. Domptail (Eds.), *Degrowth, decolonization and development: When culture meets the environment* (pp. 203–220). Cham: Springer. https://doi.org/10.1007/978-3-031-25945-6_9

Durose, C., Perry, B., & Richardson, L. (2022). Is co-production a 'good concept? Three responses. *Futures*, 142, Art. 102999, 1–11

Egbai, U., & Chimakonam, J. (2019). Why conversational thinking could be an alternative method for intercultural philosophy. *Journal of Intercultural Studies*, 40(2), 172–189. https://doi.org/10.1080/07256868.2019.1577811

Emeagwali, G. (2013). Intersections between indigenous knowledge and economic development in Africa. In A. Abdi, & E. Shizha (Eds.), *Indigenous discourse on knowledge and development in Africa* (pp. 31–45). New York: Routledge

Faimau, G. (2013). *Socio-cultural construction of recognition*. Newcastle Upon Tyne: Cambridge Scholars Publishing

Faimau, G., Sello, N., & Nkhukhu-Orlando, E. (2023). Print media coverage and the socio-contextual representation of climate cahnge in Botswana. In J. Ahrens, & E. Halbmayer (Eds.), *Climate change epistemologies in Southern Africa: Social and cultural dimensions* (pp. 149–166). Abingdon: Routledge. https://doi.org/10.4324/9781003180814-9

Gothe, J., & De Santolo, J. (2022). Decolonising design practices and research in unceded Australia: reframing design-led research methods. *Architecture_MPS*, 21(1), Art. 2. https://doi.org/10.14324/111.444.amps.2022v21i1.002

Guilherme, M. (2021). Intercutural responsibility: Critical inter-epistemic dialog and equity for sustainable development. In W. Filho, A. Azul, L. Brandli, A. Salvia, & T. Wall (Eds.), *Partnerships for the goals* (pp. 599–610). Berlin: Springer. https://doi.org/10.1007/978-3-319-71067-9_75-1

Gurevitch, Z. (1989). The power of not understanding: The meeting of conflicting identities. *Journal of Applied Behavioral Science*, 25(2), 161–173

Hambira, W., Saarinen, J., & Moses, O. (2020). Climate change policy in a world of uncertainty: changing environment, knowledge, and tourism in Botswana. *African Geographical Review*, 39(3), 252–266. https://doi.org/10.1080/19376812.2020.1719366

Hanafi, S. (2023). Toward a dialogical sociology: Presidential address – XX ISA World Congress of Sociology 2023. *International Sociology*, 1–24. https://doi.org/10.1177/02685809231199678

Hill, L., Ghorpade, S., & Galappaththi, M. (2023). Toward decolonizing sustainability research: a systematic process to guide critical reflections. *Facets*, 8, 1–11. https://doi.org/10.1139/facets-2022-0254

Huntsley, J., & Brentnall, C. (2021). Generative critical conversation: A method for developing reflexivity and criticality. *Journal of Small Business Management*, 61(6), 2788–2817. https://doi.org/10.1080/00472778.2021.1938096

Kay, N., & Gaymard, S. (2020). Climate change in the Cameroonian press: An analysis of its representations. *Public Understanding of Science*, 30(4), 417–433. https://doi.org/10.1177/0963662520976013

Ketlhoilwe, M. (2013). Improving resilience to protect women against adverse effects of climate change. *Climate and Development*, 5(2), 153–159

Landman, K., & Makakavhule, K. (2021). Decolonizing public space in South Africa: from conceptualization to actualization. *Journal of Urban Design*, 26(5), 541–555. https://doi.org/10.1080/13574809.2021.1880885

Li, Y. (2023). Intentionality, not just agency: bringing intended meaning back into the micro-macro institutionalization processes. *Culture and Organization*, 29(4), 271–297. https://doi.org/10.1080/14759551.2023.2167082

Makwatse, K., Modie, L., Mpalo, M., & Mapitsa, C. (2022). Gender and equity considerations for building climate resilience: Lessons from rural and periurban Botswana. *Sustainability*, 14(17), 10599. https://doi.org/10.3390/su141710599

Maldonado-Torres, N. (2006). Cesaire's gift and the decolonial turn. *Radical Philosophy*, 9(2), 111–138

Matolino, B. (2020). Whither epistemic decolonization. *Philosophical Papers*, 49(2), 213–231. https://doi.org/10.1080/05568641.2020.1779605

Meghji, A. (2021). *Decolonizing sociology: An introduction*. Cambridge: Polity

Mignolo, W. (2011). Geopolitics of sensing and knowing: on (de)coloniality, border thinking and epistemic disobedience. *Postcolonial Studies*, 14(3), 273–283

Mignolo, W. D. (2005). *The idea of Latin America*. Malden, MA: Blackwell Publishing

Mignolo, W. D. (2014). Further thoughts on (de) coloniality. In S. Broeck, & C. Junker (Eds.), *Postcoloniality – decoloniality – black critique* (pp. 21–52). New York: Campus Verlag

Mignolo, W. D., & Walsh, C. E. (2018). *On Decoloniality: Concepts, Analytics, Praxis*. Durham: Duke University Press. https://doi.org/10.1215/9780822371779

Mistry, J., Jafferally, D., Ingwall-King, L., & Mendonca, S. (2020). Indigenous knowledge. In A. Kobayashi (Ed.), *International encyclopedia of human geography* (pp. 211–215). Amsterdam: Elsevier

Mogomotsi, P., Skelemani, A., & Mogomotsi, G. (2020). Climate change adaptation strategies of small-scale farmers in Ngamiland East, Botswana. *Climatic Change*, 159(3), 441–460. https://doi.org/10.1007/s10584-019-02645-w

Mugari, E., Masundire, H., & Bolaane, M. (2000). Adapting to climate change in semi-arid rural areas: A case of the Limpopo basin part of Botswana. *Sustainability*, 12(20), 1–34

Ndlovu-Gatsheni, S. (2013). Why decoloniality in the 21st Century? *The Thinker*, 48, 10–15

Ndlovu-Gatsheni, S. (2015). Decoloniality as the future of Africa. *History Compass*, 13(10), 485–496. https://doi.org/10.1111/hic3.12264

Ndlovu-Gatsheni, S. (2019). Provisional notes on decolonizing research methodology and undoing its dirty history. *Journal of Developing Societies*, 35(4), 481–492. https://doi.org/10.1177/0169796x19880417

Ndlovu-Gatsheni, S. (2020). *Decolonization, Development and Knowledge in Africa: Turning Over a New Leaf*. London: Routledge. https://doi.org/10.4324/9781003030423

Ngwenya, B., Thakadu, O., Magole, L., & Chimbari, M. (2017). Memories of environmental change and local adaptations among molapo farming communities in the Okavango Delta, Botswana – A gender perspective. *Acta Tropica*, 175, 31–41. https://doi.org/10.1016/j.actatropica.2016.11.029

Nothias, T. (2020). Postcolonial reflexivity in the news industry: The case of foreign correspondents in Kenya and South Africa. *Journal of Communication*, 70(2), 245–273. https://doi.org/10.1093/joc/jqaa004

Parekh, B. (2001). *Rethinking multiculturalism: Cultural diversity and political theory*. Basingstoke: Macmillan

Perry, B. (2022). Co-production as praxis: Critique and engagement from within the university. *Methodological Innovations*, 15(3), 341–352. https://doi.org/10.1177/20597991221129773

Quijano, A. (2002). The return of the future and questions about knowledge. *Current Sociology*, *50*(1), 75–87

Quijano, A. (2007). Coloniality and modernity/rationality. *Cultural Studies*, 21(2–3), 168–178

Saarinen, J., Hambira, W. L., Atlhopheng, J., & Manwa, H. (2012). Tourism industry reaction to climate change in Kgalagadi South District, Botswana. *Development Southern Africa*, 29(2), 273–285

Saarinen, J., Moswete, N., Atlhopheng, J., & Hambira, W. (2020). Changing socio-ecologies of Kalahari: Local perceptions towards environmental change and tourism in Kgalagadi, Botswana. *Development Southern Africa*, 37(5), 855–870. https://doi.org/10.1080/0376835X.2020.1809997

Samuel, C., & Ortiz, D. (2021). 'Method and meaning': Storytelling as decolonial praxis in the psychology of racialized peoples. *New Ideas in Psychology*, 62, Art. 100868. https://doi.org/10.1016/j.newideapsych.2021.100868

Santos, B. (2014). *Epistemologies of the South: Justice against epistemicide*. London: Routledge

Smith, L. (1999). *Decolonizing methodologies: Research and indigenous peoples*. London and New York: University of Otago Press

Sweetman, B. (2008). *The vision of Gabriel Marcel: Epistemology, human person, the transcendent*. Leiden: Brill

Taylor, C. (1994). The politics of recognition. In A. Gutman (Ed.), *Multiculturalism: Examining the politics of recognition* (pp. 25–73). New Jersey: Princeton University Press

The Vision 2036 Presidential Task Team. (2016). *Vision 2036 – achieving prosperity for all*. Gaborone: Government of Botswana

Topa, W. (2022). How decolonizing/indigenizing schooling can build community: A fictional case study on bringing 'spirituality' back to learning. *International Journal of Educational Research*, 115, Art. 102018. https://doi.org/10.1016/j.ijer.2022.102018

Trisos, C., Auerbach, J., & Katti, M. (2021). Decoloniality and anti-oppressive practices for a more ethical ecology. *Nature Ecology and Evolution*, 5, 1205–1212

UNESCO. (n.d.). *Local and Indigenous Knowledge Systems (LINKS)*. UNESCO. Retrieved March 31, 2024, from https://en.unesco.org/links

von Schonfeld, K. C., Tan, W., Wiekens, C., Salet, W., & Janssen-Jansen, L. (2019). Social learning as an analytical lens for co-creative planning. *European Planning Studies*, 27(7), 1291–1313. https://doi.org/10.1080/09654313.2019.1579303

Transnational Climate Change Governance and the Postcolonial Appendices: The Case of the IPCC

8

Jörn Ahrens

Abstract

On an international scale, the Intergovernmental Panel on Climate Change (IPCC) is regarded as the most important institution providing global expertise on the global development of climate change, about which it periodically publishes reports. Founded in 1988 by the World Meteorological Organization and the United Nations Environment Programme, the IPCC's purpose is to 'provide regular assessments of the scientific basis of climate change, its impacts and future risks, and options for adaptation and mitigation' (IPCC, n.d.). This chapter examines the rationale of the IPCC focusing on its knowledge production practices, especially including possible effects on the maintenance of postcolonial structures or attitudes. What does it mean when knowledge about a phenomenon globally perceived as pressing, which, as the majority of experts agree, demands quick transnational action, is funnelled into a single pool of shared global knowledge? From a postcolonial angle the chapter reflects on the epistemological question of whether the work of the IPCC contributes to the overall establishment of a postcolonial climate change regime. It discusses the legacy of the IPCC as the leading institution of epistemological climate change governance, reflects on the IPCC's mission and its values—constructiveness, universalism of knowledge and impacts on depoliticisation.

J. Ahrens (✉)
Justus Liebig University, Giessen, Germany;
North-West University, Potchefstroom, South Africa
e-mail: Joern.Ahrens@sowi.uni-giessen.de

A. C. Terblanché-Greeff et al. (eds.), *Climate Change and the Postcolonial*, Kulturelle Figurationen: Artefakte, Praktiken, Fiktionen,
https://doi.org/10.1007/978-3-658-49328-8_8

Keywords

Climate change · IPCC · Universalism · Knowledge production · Postcolonial · Planetary

Introduction

At a press conference on climate change on 27 July 2023, the United Nations (UN) Secretary-General António Guterres chose an unambiguously dramatic tone:

> For vast parts of North America, Asia, Africa and Europe – it is a cruel summer. For the entire planet, it is a disaster. And for scientists, it is unequivocal – humans are to blame. All this is entirely consistent with predictions and repeated warnings. The only surprise is the speed of the change. Climate change is here. It is terrifying. And it is just the beginning. The era of global warming has ended; the era of global boiling has arrived. [. . .] No more hesitancy. No more excuses. No more waiting for others to move first. There is simply no more time for that. It is still possible to limit global temperature rise to 1.5 °C and avoid the very worst of climate change. But only with dramatic, immediate climate action. We have seen some progress. A robust rollout of renewables. [. . .] But none of this is going far enough or fast enough. Accelerating temperatures demand accelerated action. (Guterres, 2023)

Three years earlier, on 2 December 2020, Guterres delivered a speech at Columbia University in which he dealt, as a matter of priority, with the global challenge of climate change. Although his tone was not as dystopian as in 2023, he made his view on the condition of the planet very clear:

> To put it simply, the state of the planet is broken. [. . .] Humanity is waging war on nature. This is suicidal. Nature always strikes back – and it is already doing so with growing force and fury. [. . .] Let's be clear: human activities are at the root of our descent toward chaos. But that means human action can help solve it. Making peace with nature is the defining task of the 21st century. It must be the top, top priority for everyone, everywhere. [. . .] A new world is taking shape. More and more people are recognizing the limits of conventional yardsticks such as Gross Domestic Product, in which environmentally damaging activities can count as economic positives. Mindsets are shifting. More and more people are understanding the need for their own daily choices to reduce their carbon footprint and respect planetary boundaries. (Guterres, 2020)

It is clear from the above quote that he has been offering this line of argumentation about global climate change responsibilities for years. His tone is becoming increasingly dramatic, as adequate climate action on a global scale is obviously missing.

What sounds like a simple statement that 'for scientists, it [was] unequivocal' that climate change and its social and ecological impacts were a result of human agency and its domination over nature, depends in fact on a distinctive set of epistemological preconditions, which need to be developed and disseminated by appropriate social institutions. It is not least the Intergovernmental Panel on Climate Change (IPCC) that crucially contributes to the establishment of the broadly held notion of credibility of science that also feeds most of the political decisions that need to be taken with regard to the challenge of anthropogenic climate change. As Silke Beck argues, the publication of the IPCC's assessment reports and special reports since the early 1990s has subsequently shown 'that the scientific evidence supporting [...] two basic points is now overwhelming' (Beck, 2011: 297). These points are, firstly, the now widely accepted acknowledgement that global warming is occurring, and, secondly, 'that human activity is more likely than not a significant cause' (Beck, 2011: 297). Today, it seems only reasonable to agree with Guterres's point of view as the argument for climate change is well supported by research.

Scientific findings seem to support Guterres' position which includes an ethical as well as a scientific and political dimension. After all, the UN Secretary General receives information about the complex, global and contested phenomenon of climate change from the IPCC, its working groups and reports. In the public and political discourse alike, the reality of climate change is often considered to be self-evident. However, as has been pointed out time and again, the phenomenon is much too complex for it to be possible to base that reasonable insight on individual experience (Hulme, 2009; Jasanoff, 2010). For this reason, the perspective of the individual social actor needs to be supplemented by an abstract, evidence-based approach to knowledge production. This makes the UN secretary-general the most prominent representative of the general need for reliable data and best-informed analytical skills to put such information together and to model, on the basis of this data and analysis, a working narrative and a representative concept of the current state of global climate change.

The IPCC is broadly seen as the most important institution to provide such global expertise by periodically publishing, in particular, reports on various aspects of the global development of climate change. Founded in 1988 by the World Meteorological Organization (WMO) and the United Nations Environment Programme (UNEP), the IPCC's purpose is, according to its website, to 'provide regular assessments of the scientific basis of climate change, its impacts and future risks, and options for adaptation and mitigation' (IPCC, n.d.). The formation of the IPCC goes back to the First World Climate Conference convened by the UN World Meteorological Organization (WMO) in 1979 in Geneva, Switzerland. Conference participants did not only focus on the scientific discourse on global warming, which at the time was only presumed to exist, but also

> organized four working groups to look into climate data, the identification of climate topics, integrated impact studies, and research on climate variability and change. It called on the world's governments 'to foresee and prevent potential manmade changes in the climate that might be adverse to humanity'. (Blau, 2017: 24)

Eventually, the conference 'led to the establishment of the World Climate Programme' and to the creation [of the IPCC]' (Blau, 2017: 24). Today, the IPCC consists of 195 member states and several observing non-governmental organisations.

Against this background, the following chapter examines the rationale of the IPCC with a focus in its knowledge production practices, especially including possible impact on the maintenance of postcolonial structures or attitudes. What does it mean when knowledge about a phenomenon globally perceived as pressing, which, as the majority of experts agree, demands quick transnational action, is funnelled into a single pool of shared global knowledge? This chapter discusses whether, in epistemological terms, the IPCC contributes to the establishment of a global order of climate change that relates to postcolonial premises. Therefore, it firstly discusses the legacy of the IPCC as the leading institution of epistemological climate change governance. Secondly, it outlines three missions of the IPCC: to assess the existing knowledge on climate change, to administrate climate change and to contribute to climate governance measures. Subsequently, three essential values of the IPCC are identified—its constructiveness, the universalism of knowledge and the impacts it has on depoliticisation efforts. Thirdly, the chapter addresses the concept of planetary universalism as a new twofold global regime for climate change governance that includes climate knowledge and climate action. On the basis of the structure of climate change equity, this section argues that this regime concurrently establishes standards which are unequal to a significant extent. Finally, the chapter reflects on the IPCC's practices from a postcolonial angle. This also means that the chapter does not explicitly or solely focus on the Southern African region. Instead, its aim is to provide a more general reflection on current procedures in the implementation and institutionalisation of global climate governance.

The IPCC and Its Legacy

According to the IPCC's official website, the 'objective of the IPCC is to provide governments at all levels with scientific information that they can use to develop climate policies. IPCC reports are also a key input into international climate change negotiations. The IPCC is an organisation of governments that are members of the United Nations or WMO' (IPCC, n.d.). In 1990, it provided its First Assessment Report (AR1), which was followed by a new assessment report every five to six years. Since the 2000s, and especially since the 2015 Paris Agreement, the IPCC has not only substantially increased its production of special reports but has also continuously increased its influence on how climate change is perceived in the global discourse. Today, it is widely accepted as the 'authoritative voice of international science' (Beck and Mahony, 2018: 1) in the production of climate knowledge and has published its Sixth Assessment Report (AR6) in March 2023. The IPCC does not conduct its own research but relies on published research, which is evaluated and put together by a wide range of experts working with the IPCC. Therefore, 'an open and transparent review by experts and governments around the world is an essential part of the IPCC process, to ensure an objective and complete assessment and to reflect a diverse range of views and expertise. Through its assessments, the IPCC identifies the strength of scientific agreement in different areas and indicates where further research is needed' (IPCC, n.d.).

Today, the IPCC reports are seen internationally as the 'gold-standard' of climate change science as the research on climate change and its manifold global consequences that goes into these reports is considered to be of utmost quality. As Alix Kashdan and Kristyn Ostanek put it in an article for the United Nations Foundation, the IPCC's 'gold-standard climate science reports inform debate and shape policymaking' (Kashdan and Ostanek, 2022) as they impact the 'global ambition' in efforts to slow down and limit global warming. Hence, the IPCC's mission is to provide exactly that sort of reliable expertise on the development of global warming that forms the basis of statements like those quoted from the UN secretary-general's speech. Politics, economy and civil society worldwide draw information on which to base potential adaptation and mitigation measures from these reports that also provide information on topics that require further research: 'The IPCC provides policymakers with regular scientific assessments on climate change, its impacts and risks, as well as adaptation and mitigation options' (IPCC, 2019: VII). The resulting interface of purely scientific assessment and policy consultation has repeatedly been the subject of debate and critique. However, according to Beck (2009), the IPCC's unique success as an international institution is not least due to the fact that, from its beginnings, the IPCC managed to speak with one voice and in the name of international science. Only by using this strategy, she argues, did it succeed in bridging the gap between science and politics and providing a reliable basis for international politics (Beck, 2009).

In line with its purpose as described above, one major mission of the IPCC is to provide global, consistent and unequivocal knowledge about climate change. At least since the approval of the 2015 Paris Agreement, climate change has been acknowledged as a pressing global challenge that needs to be confronted transnationally through co-ordinated action of multiple actors worldwide. The absolute precondition to make this possible is to establish a shared global understanding that provides a frame for the contemporary global warming status quo. This is rather demanding and requires the standardisation of diverse global epistemologies into a unilateral and universalist approach. From this perspective, as Arturo Escobar observes, forms of knowledge that would take into account a plurality of 'relational worlds' were possibly more 'appropriate to the conjuncture of modern problems without modern solutions' than was the case with their academic counterparts in the Western sciences (Escobar, 2020: 49).

Escobar emphasises that such knowledge was

> located within a twofold context: that of the need for civilizational transitions, on the one hand, and the planetary dynamics brought to the fore by global climate change, the destruction of biodiversity, and the Anthropocene, on the other hand. (Escobar, 2020: 49)

He thus refers to an essential 'diversity of forms of knowledge held by those whose experiences can no longer be rendered legible by Eurocentric knowledge in the academic mode, if they ever were' (Escobar, 2020: 41). What Escobar is aiming at could therefore be called a knowledge *pluriverse*. That is, a heterotopical realm that consists of a plurality of cultural ontologies and epistemological practices that may be best suitable for generating adequate responses to the present global condition, including climate change. Other than this, the unilateralisation of climate change as promoted by the IPCC is part of a process of 'epistemological monopolization' (Sousa Santos, 2018a: 161) initiated by modern science that results in the suppression of knowledge from the Global South. Although it is especially focused on the way climate change affects the whole planet, the demand for global responses to climate change abandons polycentric approaches to knowledge and establishes instead one master approach:

> The epistemologies of the North have contributed crucially to converting the scientific knowledge developed in the global North into the hegemonic way of representing the world as one's own and of transforming it according to one's own needs and aspirations. (Sousa Santos, 2018b: 6)

In this sense, the IPCC needs to be regarded as a major actor in building one unilateral global epistemology. Therefore, it can be said to hold an indexicality of postcolonial domination. Global measures for climate change governance are contingent on common knowledge that is shared by diverse actors from politics to economy to everyday practices. Hence, a streamlined understanding of the problem is indispensable in making successful engagement with a global challenge like climate change possible. This is supposed to guarantee equal access to knowledge about the phenomenon and its impacts on the social condition to all social actors, irrespective of their geographical or social position. Therefore, it is necessary firstly to create a common basis on which to make transnational action and its collaborative co-ordination possible. Procedures of knowledge production, knowledge dissemination and information analyses are most important in this regard, as co-operative global measures for climate change governance demand the provision of such consensual knowledge.

Predicting the impacts of global climate change is only possible if we accept the epistemological claim that the concept of climate change has been successfully turned into a valid global 'social imaginary' (Taylor, 2004). If, according to Charles Taylor, the social imaginary 'extends beyond the immediate background understanding that makes sense of our particular practices' (Taylor, 2004: 25). The social imaginary is what is added to the basic knowledge and experience of lifeworldly interaction and what provides meaning for practices that elude immediate understanding. As such, it fabricates meaning, even of concepts as abstract as climate change, which cannot possibly be experienced on the lifeworld level. The experience of climate change is only possible on the abstract level of epistemological recognition that is defined globally per se. As such, it has now become an essential part of the 'ensemble of imaginings that enable our practices by making sense of them' (Taylor, 2004: 165). Climate change needs to be translated into effective social imaginaries; as a global phenomenon of far-reaching consequences, it even develops into one of the major social imaginaries of our time. However, due to its precarious status as a concept of global significance that at the same time has hazardous impact on the same societies it generates meaning for. Hence, as a social imaginary, climate change as much a contributor to the social embeddedness of social actors in the Anthropocene as it is a major force for the contemporary disembedding of social institutions due to the environmental crises to which it gives rise (Taylor, 2004).

In this situation, knowledge about the origins and causes of climate change is on the one hand shaped by continuous and co-ordinated national political endeavours, but seeing that it is a planetary challenge it is even more shaped by transnational legal and political frameworks. On the other hand, it refers to the rhizomatic practices of global research. These practices vary according to the particular climate change conditions with which societies are confronted, the different discourse constellations about climate change in these societies and different resource allocation patterns. Being one of the most important actors in the production and provision of a globally effective social imaginary of climate change, the expertise provided by the IPCC offers institutionalised and evaluated knowledge that does not only provide knowledge about this menacing environmental phenomenon, but also epistemological safeguarding. With its reports the IPCC makes the status quo of climate change and its prospective development almost uniquely plausible planetwide. However, by constantly offsetting bad news about how far climate change has already developed by assurances that there is still room for global agency, the IPCC also brings hope. Given its unique and central position in the global discourse, this global epistemology serves as a new sovereign of reason as its comprehensive knowledge monitoring is enabling responses to climate change (Tol, 2011; Livingston and Rummukainen, 2020, 2023).

Against this background, the implementation of the IPCC as the global 'gold-standard' institution for the assessment of climate change knowledge also requires techniques to streamline this knowledge through review practices. Via these evaluation techniques, climate change knowledge is declared decisive, and its global dissemination is guaranteed. Clark Miller concludes that such work of elaboration and reinforcement of a 'global, systemic understanding of climate and climate change' makes an important contribution to the design of a 'vision of natural order that made clear the necessity for, and possibility of, a global politics of climate' (Miller, 2004: 55). The legacy of a new global natural order is thus based on solid information that argues for the necessity and adequacy of such an order. At the same time, this 'natural order'—of which even an institution like the IPCC with its unquestionable ethical and conservationist agenda is an important part—still integrates itself into the history of domination over nature in modernity. As Walter Mignolo argues, nature 'became a repository of objectified [...] materiality that existed for the fulfilment of the economic goals of the 'masters' of the materials' (Mignolo, 2011: 13). This transformation of nature from alter ego into natural resources has been regarded a 'sign of progress and modernization'

(Mignolo, 2011: 13). Following the successfully established 'Western cosmology' within an order of colonial domination, he argues, 'Western knowledge became a commodity of exportation of the modernization of the non-Western world' (Mignolo, 2011: 13). It is this agenda of domination over nature and the human condition which regards both as resources for consumption that furnishes the project of global colonisation as a 'process and project of Westernization' (Mignolo, 2011: 32) that also allows for the 'control and management of knowledge and subjectivities' (Mignolo, 2011: 33). The domination over nature also proliferates the domination of Western epistemologies and the establishment of epistemological power.

Today, the IPCC serves as the key institution for maintaining climate change knowledge and still relies on precisely this concept of domination and definition of nature and knowledge. It is exactly this understanding of nature as a resource and a commodity that has led to the current anthropogenic climate change crisis, and it shapes the IPCC's approach to the crisis as well as possible adaptation or mitigation measures, which is less recognised in discourse. As the IPCC engages in linking knowledge and power, Miller argues, it contributes to the establishment of a level of discourse power that transcends the regular rationality of politics and replaces it with a representation of 'climate as a global-scale natural system' (Miller, 2004: 55). Thus, the IPCC has managed to establish a new kind of transcendent political order beyond the established political structures. The ethos of global climate change responsibilities that refer to the scientifically grounded evaluation of the status quo of global warming serves as a normative meta-order for the planetary condition that, as a new variant of the one-world order, strictly ignores global differences in positioning, affectedness or interest. The IPCC serves as both a medium for reliable communication and an authority for knowledge assessment. The IPCC, as Miller concludes, articulates 'a narrative of global politics in which experts play a powerful role as politically neutral agents' (Miller, 2004: 61). Hence, the establishment of the IPCC has created an instance for interpretation that is essentially political, but which the IPCC addresses only indirectly as it does not have a political mandate and does not want to have it.

The IPCC's Missions

Beyond its above-mentioned basic agenda to 'provide regular assessments of the scientific basis of climate change' (IPCC, n.d.), the IPCC pursues three missions. The first is to review the existing climate change knowledge and pass academic judgment on it. The success of the IPCC stems not least from the fact that it formally follows the agenda of systematic and global knowledge assessment, although by doing this it eventually creates, arranges and interprets, albeit subtly, exactly the knowledge that will help proliferate the international discourse on climate change. To assess such knowledge, the IPCC first needs to collect it. As Beck and Mahony rightly emphasise, one of IPCC's concerns is to maintain its established perception in the public sphere which constitutes its 'independence as an assessor, rather than a producer, of knowledge' (Beck and Mahony, 2018: 7). However, this demand eventually runs counter to the IPCC's de facto core competence, which is the production of a particular form of newly established environmental and societal knowledge which has only been made possible by the IPCC's systematic and strategic assessment practices.

Hence, referring to the concept developed by Fleck (1980), the IPCC establishes a worldwide 'thought collective' that makes it possible to establish climate change as a global scientific fact. This is important in the process of establishing a globally shared awareness of climate change not only as an environmental anomaly but as fundamental challenge to existing societies. When climate change consolidates worldwide as a scientific fact, the incidental epistemology will also successfully establish itself as a form of unilateral knowledge about the global condition. Therefore, it counteracts postcolonial endeavours for a pluralisation of knowledge. When Sousa Santos outlines that the epistemologies of the South consist of an 'infinite plurality of finite ways of knowing human experience in the world' (Sousa Santos, 2016: 23) as he strives to establish an 'epistemological diversity of the world' (Sousa Santos, 2016: 24), an epistemology of climate change as the common planetary situation precludes such diversity and demands a unilateral perspective instead. This is the knowledge which, as the 'gold-standard', defines the global reality of climate change and the needs of nature in this process. According to Beck, the IPCC is not even aware of its essential political quality but insists on its 'belief in the neutrality of science' (Beck, 2011: 299), particularly to avoid its own politicisation and to 'preserve its scientific credibility' (Beck, 2011: 299). Indeed, it may be the case that this analysis is correct and that the IPCC's concerns about its own scientific reputation repress the sense of its global political influence and agency.

However, Beck also argues that the IPCC obviously has an 'awareness about political terrain' (Beck, 2011: 299), even if, at the same time, it avoids the latter. In fact, it is this adversity to politicisation that makes the political capital of the IPCC and guarantees its influence. Only by remaining a politically neutral, scientifically objective actor in the global climate change discourse, the IPCC also manages to remain the unquestioned technocratic expert on climate change, the most important political consultant within the transnational discourse and among the bodies of climate change governance. It is its agenda of being unbiased or non-political that makes the political quality of the IPCC. To maintain this quality of technocratic neutrality, it follows a specific set of rules when, with few exceptions, it considers only peer-reviewed academic work that has mostly been published in international English language journals. Hence, the first mission of the IPCC is to filter down the highly complex state of international research. Only what is judged by the IPCC working groups to be of sufficient quality contributes to the constitution of the generally acknowledged scientific basis of climate change that also provides the effective order of discourse on a global scale.

What is eventually presented and assessed as the basis of climate change knowledge is itself based on a selection that, as has been repeatedly shown (Hulme and Mahony, 2010; Beck, 2011; Beck and Mahony, 2018; Collyer, 2018), prioritises contributions from the natural sciences over those from the social sciences and privileges contributions from the Global North over the Global South, especially since the IPCC's approach tends to streamline knowledge and exclude or marginalise dissenting forms of knowledge. The underlying 'privileged validity of modern science' has been 'developed predominantly in the Global North' (Sousa Santos, 2018b: 5). On the coloniality of this knowledge, seeing as it is also part of the emergence of transnational climate change knowledge and governance, Mignolo stresses that it would not

> mean that knowledge was colonized, but that hegemonic ways of knowing and disciplinary world-making, since the European Renaissance, were instruments of colonization and, as a consequence, of colonization of non-European knowledge. (Mignolo, 2011: 189)

As Reiner Grundmann and Nico Stehr observed, the issue of climate change 'was from the beginning the domain of natural scientists' (Grundmann and Stehr, 2010: 900), whereas, as they argue, the social sciences have not only been marginalised in the discourse, but for a long time they did not even engage in the field (Grundmann and Stehr, 2010). The first IPCC Assessment Report (AR1) was not only rather strict in following this assessment agenda, it also almost exclusively involved scientists from the Global North. Both the limitation to a seemingly neutral practice of assessment and the widespread absence of scientists from the Global South drew criticism as several developing countries expressed their dissatisfaction. Despite improvements, the situation did not change substantially (Beck and Mahony, 2018). At the same time, the 'incorporation of 'grey literature' sources'—which means information that is produced outside of traditional academic publishing and distribution channels, often also not well represented in indexing databases—remains a problem for the IPCC and its approach as some regions are not adequately covered by recognised academic research, thus the 'the global centers of climate change research' become over-represented (Beck and Mahony, 2018: 5).

The aforementioned feeds the developing countries' 'skepticism toward the IPCC's "one world" vision of climate change' (Miller, 2004: 62). Efforts made by the IPCC to close these gaps still cannot be regarded as effective. Its second mission, one that makes the IPCC an important body in the administration of climate change knowledge, stems from its task of curating global climate change knowledge. Regular reporting on either special problems or the assessment of global climate change knowledge constitutes an administrative practice. By applying this practice, the IPCC not only authoritatively selects valuable knowledge contributions but also translates them into voluminous and well-structured reports where this knowledge is organised in a distinctive order and forms a specific narrative and meaning. The administrative processing and editing of the content made available in this way concurrently creates specific modes of meaning related to the global state of climate change and possible measures to approach it.

Finally, this implies the third mission of the IPCC, which establishes it as an important actor in the attempt to implement globally effective climate governance within what is today known as governance measures for a sustainable transformation of the economy and societies worldwide. Such governance demands a solid epistemological basis of valuable and widely acknowledged data on which any collaborative action is based. All transnational bodies for the establishment of global climate change governance appeal to the authority of the IPCC's reports on climate change. This involves most prominently the negotiations and declarations of the annual UN-run Conference of Parties (COP), the drafting of National Development Plans (NDPs) which are to be implemented according to the 2015 Paris Agreement by all signatories, or any reasoning about the meaning and impact of climate change by governments, economic operators and social movements (Jamison, 2010; Boykoff and Pearman, 2019; Sanford et al., 2021; Hartz, 2022; Koch et al., 2023). Hence, the IPCC wields substantial influence on the production of the global reality of climate change as it represents the key institution that selects the available knowledge about climate change, organises it and furnishes it with globally acknowledged meaning.

All of these measures aim at a transition from an understanding of the global to the understanding of the planetary. The latter, however, also lends itself to critical analyses, such as when Escobar calls the human species a 'disruptive force on the planet earth' (Escobar, 2020: 52). Next to 'energy, poverty, and meaning', climate, he argues, is one of those crises of the contemporary that result from a broader 'crisis of the Western "civilizational model"' (Escobar, 2020: 51). According to Escobar, this crisis creates new opportunities for many social movements, including 'ethnic movements' or 'peasant networks', now provided with new means for critique as well as with options for post-extractivism agendas (Escobar, 2020: 51). Escobar's assessment is clearly not shared by Dipesh Chakrabarty, who, with the human species 'becoming a geo-physical force on the planet', sees a new quality of collective social existence emerging (Chakrabarty, 2012: 13). Such planetary existence, then, clearly lacks an 'ontological dimension' as 'our thinking about ourselves now stretches our capacity for interpretive understanding' that requires 'nonontological ways of thinking the human' (Escobar, 2020: 51). At the same time the planetary 'geological agency of humans', Chakrabarty argues, needs to be connected with 'the way humans interfere in the distribution of natural life on the planet' (Chakrabarty, 2016: 106). Contemporary ecological crises like climate change therefore express a threat 'not [to] the geological planet itself but [to] the very conditions, both biological and geological, on which the survival of human life as developed in the Holocene period depends' (Chakrabarty, 2009: 213). In this perspective, the planetary, although it gives expression to the new situation of the human species' geo-political influence, still detaches from a more classical understanding of political struggle. The IPCC seeks to respond to this dimension of the planetary that realises that contemporary crises, especially ecological ones, need to integrate not only the human species, but include any form of life on the planet. The planetary, in this regard, is an existential category for the survival of the entirety of life on the planet in its current shape.

Due to its above-outlined agenda of knowledge assessment instead of knowledge production, the IPCC first and foremost refers to the hard sciences as disciplines that are thought to generate objective data when it puts together its assessment reports. One effect of this procedure is that the IPCC establishes itself as the leading institution in the social construction of climate change. As both the provider and administrator of global climate change knowledge, it is the central actor of a globally performative epistemology for 'the assessment, validation and mobilization of such knowledge claims about climate change' (Hulme and Mahoney, 2010: 2). Contrary to positions that insist on a robust scientific consensus about the nature of climate change, Grundmann and Stehr insist on the 'virtues of constructivism' (Grundmann and Stehr, 2010: 898). Steven Yearley points out that knowledge about climate change and the future climate is nothing less than 'undeniably (and uncontroversially) constructed' (Yearley, 2009: 390). Firstly, he argues that this knowledge was based on 'projections of the future behaviour of an enormously complex system' (Yearley, 2009: 390) as the scenarios, also applied by the IPCC, were based on mere prediction. Secondly, Yearley states that these projections would depend on assumptions about how social and political actors will act (Yearley, 2009). Hence, the assessment of available scientific data not only results in the construction of the present reality of climate change but, above all, yields the prediction of its future evolvement.

Beyond this analysis of the IPCC's formal approach as effectively constructing the social reality of climate change, Miller also sees a strategic constructivism applied by the IPCC. He argues that through its reports, based on careful selection of scientific results thus made assessable, the IPCC would deliberately construct 'a discourse that framed climate change as a risk to the global environment' (Miller, 2004: 47). The preceding analysis stresses the greater political meaning of the IPCC in the devising of strategies for global governance of climate change as it provides the concepts and keywords while at the same time remaining an officially unpolitical, scientific actor due to its agenda of ideological neutrality. The IPCC is an instructive example for a political governance board that consists of experts not considered governance actors. As early as 2010, Ulrich Beck remarked, aiming at the IPCC, that the 'discourse on climate politics so far is an expert and elitist discourse in which peoples, societies, citizens, workers, voters and their interests, views and voices are very much neglected' (Beck, 2010: 255).

Having received criticism of its first Assessment Reports, the IPCC now goes to great lengths to reflect, whenever possible, not only on all affected regions but, in some cases, also integrates 'grey literature' and, to an extent, strives to consider indigenous knowledge. The latter is seen as part of existing 'diverse forms of knowledge such as Indigenous knowledge, local knowledge and scientific knowledge' (IPCC, 2022a: 36). According to the institution's own credo, 'thousands of people from all over the world contribute to the work of the IPCC. For the Assessment Reports, experts volunteer their time as IPCC authors to assess the thousands of scientific papers published each year to provide a comprehensive summary of what is known about the drivers of climate change, its impacts and future risks, and how adaptation and mitigation can reduce those risks' (IPCC, n. d.). Hence, IPCC reports are organised as a collaborative and transnational endeavour that, in its efforts to construct and frame the globally acknowledged climate change discourse, is also expected to anticipate the required ethos and collaboration that is to inform the future climate action of national governments, transnational organisations and institutions and international as well as national NGOs.

The ethical approach to planetary responsibility that is linked to the IPCC's mission is decidedly universalist in that it suggests a state of global equity of all actors engaging with the emergence of climate change and any collateral impacts. Beck and her co-authors argue that the understanding of climate change as a 'universal global risk' results from practices of epistemological framing that '[reinforce] the assumption that more and better consensual decision support will lead to public trust and political action' (Beck et al., 2014: 82). Their analysis remains sceptical about this approach as it would greatly contribute to 'political inertia'; the strategy for framing the challenge of climate change, they say, would have 'rendered climate policy ineffective and has foreclosed the possibility of public consent' (Beck et al., 2014: 82). Beck and her co-authors are concerned that the framing of climate change knowledge and its consequences would limit climate agency to those actors that are already established in the existing transnational bodies—predominantly the national governments. Hence, alternative actors would not have a voice in the global climate agenda. However, their point is that if

> climate change risks were framed differently, then different forms of political action would open up – in relation, for example, to regional adaptation, local air quality, and energy services for the poor. (Beck et al., 2014: 82)

The assumption of the universality of climate change, as represented by the IPCC, is based on the premise that, as a planetary challenge, climate change equally impacts not only humans, but all living beings everywhere. As a planetary threat, climate change is neither geographically nor temporally limited but affects the whole of planet Earth (Chakrabarty, 2009). Such a perspective is not only expressed in statements like the aforementioned quotes by UN Secretary-General Guterres. As an historian, Chakrabarty points in that same direction when he posits humanity as the appropriate category to address the contemporary challenges of the Anthropocene, the 'placeholder for an emergent, new universal history of humans that flashes up in the moment of the danger that is climate change' (Chakrabarty, 2009: 221). However, this remains ambivalent as Chakrabarty cautions that although this universal that emerges as climate change could never be understood, it also 'poses for us a question of a human collectivity, an us, pointing to a figure of the universal that escapes our capacity to experience the world' (Chakrabarty, 2009: 222). Hence, the claim of the universality of climate change thus implies that climate change is not the only thing that concerns humans as a species, that there is also the experience of global decay that emerges from it. 'It is more like a universal that arises from a shared sense of a catastrophe' (Chakrabarty, 2009: 222). According to Chakrabarty, to speak of a 'planetary climate system' in the Anthropocene is to establish 'a totalising picture of the planet that is, at the same time, withdrawn from and inaccessible to earthlings like humans' (Chakrabarty, 2018: 265). This, he argues, would make 'a new planetary real' (Chakrabarty, 2018: 265).

Today the 'planetary' has become an influential code-word for the complete colonisation of planet Earth by the emerging dystopian future of climate change. According to Chakrabarty, the experience of climate change not only points at a future that we as humans can neither visualise nor imagine (Chakrabarty, 2009) but is also rooted in the geological dimension of a 'deep history of humanity' (Chakrabarty, 2009: 213). In this ecological and social crisis, the effects of long-term activities in history make up a gestalt that affects the whole planetary condition (not just that of a specific area or era) and 'constitutes a crisis for humans' (Chakrabarty, 2009: 213) and for non-human entities alike. Chakrabarty's point is, ultimately, that in this planetary crisis present struggles ought to be transcended. According to him, even if the global system of capitalism is a decisive driver of climate change, the important question in this situation would not be about responsibilities, but about affectedness, as the whole planet, and therefore the whole human species, is equally affected by the emergence of climate change as an unintended result of human actions in the domination of nature. In the wake of climate change, Chakrabarty argues, the species-based approach to the planetary is the only appropriate one.

Planetary Universalism

The understanding of climate change as a planetary phenomenon and problem fits the practice of representation as applied by the IPCC. Regardless of its legacy, one of its first and most important results is a consequent depoliticisation of climate change responsibilities due to its universalisation. The planetarisation of global warming abandons historical, social or territorial aspects. The planetary approach can only be one that neglects difference by adopting a meta-perspective that melts down any aspect of *otherness*. This may be advantageous as long as it is aimed at remedying inequalities, but it ignores responsibilities and contextual specificities and constellations in cases like the unequivocal climate change. Instead, the planetary perspective furnishes the phenomenon of climate change with a transcendental quality that eventually establishes it as a natural phenomenon that spreads the globe, when in fact it not only emerges from human action but is also historically and socially bound. According to Chakrabarty, it is the IPCC reports that 'burst in on us' (Chakrabarty, 2012: 9) and foster broad awareness of climate change. These reports and their practices of knowledge assessment represent climate change as a materiality, as something concrete, when previously it was in most cases perceived as abstract information and mere weather. Owing to these reports and their extensive media coverage, the dimension of global warming becomes understandable on a lifeworld level. It therefore turns into a sort of cultural experience, knowledge about the world that can no longer be neglected.

At this point, the IPCC reports resemble what Michael Hardt and Antonio Negri call the 'concrete universal' (Hardt and Negri, 2002: 35), a quality of the global order of power within 20th century modernity that Hardt and Negri term 'empire'. The 'concrete universal' transfers local specificities and their diverse negotiations into a concrete universalism of values, ethics and justice that are consequently becoming part of one equivocal global condition of power and law. Nevertheless, Hardt and Negri argue that, as effectual as the emergence of the empire may be, it clearly emerges from a state of crisis (Hardt and Negri, 2002). Likewise, the epistemological and transnational empire of climate change is an empire of crisis. Hardt and Negri's argument is that, in the 20th century, the claim for universalism has been one important element in the building of that global order of power which they call the empire. Thus, universalism is taking an ambivalent position as it stands for the acknowledgment of ethical virtues worldwide as much as it establishes a global order that follows the purpose of epistemological as well as material suppression.

This approach to universalism, which is marked by significant difference is obviously based on the legacy of poststructuralist philosophy; it also informs Achille Mbembe's approach to the postcolonial. He, too, states that there is 'only one world' that consists of the difference of 'a totality of a thousand parts' (Mbembe, 2017: 180). Therefore, humanity is only considered as a whole that remains fragile as it 'delegates itself in the world' (Mbembe, 2017: 180). According to Mbembe, by being in the world, humanity needs to integrate into the 'totality of the living world' (Mbembe, 2017: 180). Therefore, it is necessary to recognise not only the 'world of humans' but also that of nonhumans (Mbembe, 2017: 180). From this position Mbembe eventually makes his ethical claim for a universalism of humanity that derives from an experience of difference: 'The desire for the fullness of humanity is something we all share' (Mbembe, 2017: 182). That fullness would be the invalidation of inequalities—a social condition in which, as Theodor W. Adorno once put it, being different would be possible without anxiety (Adorno, 1997: 116), but in which material reciprocity and compensation for injustice would also be guaranteed. It is for this reason that Mbembe specifies his concept of reparation in the postcolonial context that, to him, is closely connected with the reciprocal validation of humanity.

The introduction of reparation would thus mean setting off a process to relaunch 'the forms of reciprocity without which there can be no progress for humanity' (Mbembe, 2017: 182). In line with Chakrabarty's position, this means that a universalism that meets contemporary needs must include not only the human species but also non-human forms of life. According to him, this is due to the fact that the anthropogenic nature of global climate change causes the collapse of the 'age-old humanist distinction between natural history and human history' (Chakrabarty, 2009: 201) that parallels the dichotomy between nature and culture that is constitutive of modern societies. As with climate change, natural and human history merge, the categorical distinction between nature and culture, between the human species and non-human life-forms, is brought into question. This tension between nature and anthropocentrically organised society is constitutive of modernity. Modern sociation consequently implies the exclusion, reification and domination over nature. As nature today becomes one of the most significant factors within the process of sociation, and also the most severe threat to the further existence not only of the human species but of the planetary condition in its current state, a fundamentally new approach in the engagement with nature seems to be required from society and its institutions and actors.

Therefore, a pluralisation of the meaning of the universal is needed especially due to the emergence of climate change and its engendering of a planetary dimension of the contemporary socio-ecological condition. This would entail expanding the universal beyond the human species to meet the requirements of an emerging planetary condition that integrates non-human forms of life. Instead, the present-day transnational agenda, as represented by the IPCC's 'gold-standard' of knowledge assessment, consequently focuses on the human species alone. In the IPCC reports, the planetary remains a medium for anthropocentric concern. A reflection on the question of what the established planetary condition of universal climate change means for an understanding of the basic categories we are still dealing with is not part of the assessment reports agenda. Whenever the IPCC approaches tropes like nature, the environment, governance, mitigation, adaptation, socio-political actors such as the human species, states or national governments, it refers to long-established concepts that rely on assumptions made with regard to the condition of classical modernity. For instance, nature is not a category in the IPCC reports, the word itself is only used in its adjectival form, 'natural'. As such, nature is approached as a resource (IPCC, 2007, 2014). It is this epistemological angle that informs the IPCC's information assessment and knowledge production. As the 'planetary' has in recent years emerged as a new umbrella category to approach a global situation beyond globalism (Shoshitaishvili, 2022), new concepts are required to gain an adequate understanding of present-day confrontations exposed by climate change. Such new framing of categories is exactly what Chakrabarty and Mbembe intend to provide from the angle of deep history and postcolonial thinking. Although Chakrabarty seems to dismantle the postcolonial critique of material inequalities in his approach in favour of laying stress on the planetary meta-crisis, both provide significant topical understanding of the universalism of difference as appropriate to our time.

In its reports, the IPCC refers to the human species as such as the supreme actor. In so doing, it effectively suppresses the plurality of differences between various actors that populate the planetary and provide global agency, and it blends the existing differences within the species into a verifiable yet inconclusive narrative of the equal affectedness of the world (Jazeel, 2011; Galaz et al., 2012). This is precisely the narrative developed by the IPCC, the story of the endangered planet and the human species, which, with every report, increases the urgency of climate action. Contrary to the approaches outlined by Chakrabarty or Mbembe, the IPCC reduces the planetary to its natural and scientific dimensions as an objective phenomenon. This approach eventually allows the IPCC, despite its otherwise ambitious ethical approach based on the universality of human affectedness by climate change, to consequently ignore the responsibility for its emergence which varies geographically, and therefore also the responsibility for climate justice (Jacobson, 2018; Vanderheiden, 2008). At this point Chakrabarty deviates from most positions that are part of the postcolonial agenda. Regarding the acuteness of climate change as a global challenge, he argues that in the wake of the new universalism of climate change, 'questions of intrahuman justice' (Chakrabarty, 2012: 14) become less important. These questions would aim at socio-political injustice and reparation itself. Instead, he says, the planetary poses new climate justice questions within the anthropogenic global order. As for the emergence of the planetary, he argues, a new mode of human existence is taking over, namely the 'collective mode of existence' of the human species as geophysical force that transcends the 'deeply human concern' of 'intrahuman justice' and further develops into an ontological mode which is necessarily 'justice-blind' (Chakrabarty, 2012: 14):

> Humans today are not only the dominant species on the planet, they also collectively constitute [. . .] a geological force that determines the climate of the planet much to the detriment of civilization itself. Today, it is precisely the 'survival of the species' on a 'world-wide scale' that is largely in question. All progressive political thought, including postcolonial criticism, will have to register this profound change in the human condition. (Chakrabarty, 2012: 15)

What is thus important for the planetary as a universal is the aspect of collectivity. The planetary as a universal does not only manifest itself as an ethical claim but demands new forms of social organisation impregnated by an understanding of collectivity, but also calls for new ways for collective interaction between humans and nonhuman life-forms. It is only this dimension of collectivity that eventually leads to new qualities of social cohesion and resilience. Chakrabarty's position is in line with the IPCC's argumentation in that he claims that in the exceptional present-day situation that upgrades the human species into a geophysical force and constitutes the era of the Anthropocene, the existence of species and the planetary gains priority at the expense of the questions of social and global justice.

The difficulty in approaching the demands of the planetary crisis via the idea of species-responsibility, as promulgated by the IPCC and—much more elaborately—by Chakrabarty, is that this requires not only the existence but also a broad acknowledgement of a global ethic of awareness of and affectedness by climate change as the new driving force for universalism. Simultaneously, the validity of such an ethic is based on the assumption that the data-turned-global-facts—not least due to the global reputation of the IPCC as an institution—is strictly, scientifically neutral. This agenda of neutrality limits the IPCC reports to the mere coverage and elaboration of the unconditional development of climate change effects with a focus on particular topics and regions worldwide. The reports only occasionally state what is missing in the respective topics and regions, but they do not elaborate on who is to accomplish this and how, and moreover, which structural condition has caused the current situation of planetary decay. Hence, the key characteristic of the IPCC's work is its focus on the factual description of how climate change develops worldwide; its achievement is thus a sort of phenomenology of the process which can be traced by the various IPCC reports and special reports. Of central importance for the presentation of the results is that in its approach and self-understanding the IPCC sticks to one of the basic prerequisites of modernity, which is the dichotomy between nature and anthropocentrically organised society. The IPCC and Chakrabarty take different positions regarding this point. In no way does the IPCC anticipate the conceptual consequences of a universalism of the natural condition that is based on the category of the human species. It addresses nature as a resource for further adaptation and management to be used to facilitate the existence of the human species and, therefore, speaks of 'natural and human systems' (IPCC, 2014: 1005).

Having adopted the influential tropes of the planetary and species universalism, the IPCC integrates them into the existing social ontologies of the order of globality (Sleeman et al., 2018; Balbi et al., 2022). This makes it possible for the IPCC to take an ethical position that responds to urgent calls for climate action like that of the UN secretary-general, but at the same time it does not reflect on the political questions of causation, social concerns about responsibility, or cultural requirements of acknowledgment (Keller, 2010; Robinson, 2020). Such neutral approach results in the neglect of the material dimension of climate change-induced inequalities (Islam and Winkel, 2017; Markkanen and Anger-Kraavi, 2019). Although in its reports the IPCC agrees that the worsening climate change proliferates and increases social inequality (IPCC, 2018, 2022b; Dasgupta, 2023), it neither makes a case for this connection systematically nor does it focus on the resulting political dimension. It focuses exclusively on climate change as a completely natural process, even if it is caused by human intervention into nature. This approach eventually implies a shift of the meaning of the political away from its established concept of modern partisan actors towards a new framing of the political that is based on a scientific expertise concerning the contemporary matters of nature. The postcolonial dimension here is that, on the one hand, the political as a realm in which climate change-vulnerable actors were able to act is denied, while on the other hand, a new dimension of the political is established that systematically excludes this material dimension. 'Global climate change reinforces disparate economic, social, and racial conditions that developed, fostered, and thrived throughout the long history of colonization, inscribing climate change as a new, slow form of imperialism that is retracing the pathways that colonialism and globalism have already formed' (Hartnett, 2021: 39).

Evidently, the work of the IPCC is of great importance for the production of a global discourse and epistemology of climate change as it remains the key transnational institution dedicated to the constant evaluation of climate change developments worldwide. This is why Ulrich Beck and Bruno Latour, in an interview with Sabine Selchow, call it a 'sort of world parliament of climate' (Selchow, 2014), as it is at the same time universal and 'speaking in the name of a science from nowhere' (Selchow, 2014). So far, the IPCC has established a twofold global regime for the governance of knowledge. This includes, firstly, the application of the trope of the planetary to foster a sense of urgency of climate change adaptation and mitigation measures. This claim to planetary universalism fails as its basic assumption of global equity in affectedness ignores the major differences in vulnerability between different parts of the world and turns out to be a contemporary adjustment to the established order of the empire of globalism. Secondly, being reasonably able to claim a planetary dimension of climate change requires the creation of a heterotopical knowledge space that would be expected to completely represent the assessable knowledge about climate change in all its varieties. This kind of a pluralisation of epistemological approaches would then provide the basis to facilitate transnational approaches for the emergence of global institutions. The ethicisation of climate change and its impacts favoured by the IPCC, however subtle, aims at the unilateralisation of climate change knowledge, which thus needs to be uniform everywhere. Scientific discourse regimes are established under this strategy. They guarantee, as Yearly observes, 'uncontroversial' construction of knowledge about climate change (Yearley, 2009). Only these measures can ensure the quality of 'humans as future-makers and of futures as cultural facts' (Appadurai, 2013: 285).

Practices

What the IPCC reports provide is not just dystopian messages about the failed climate action missions, as Guterres suggests. Rather, they are key contributions to the production of an 'ethics of possibility and [an] ethics of probability' (Appadurai, 2013: 295). What the reports achieve is the inclusion, the representation of the likely development of global climate change in all its aspects and based on the available. On the other hand, they also indicate which measures should be applied to achieve resilience or successful adaptation with regard to the UN 1.5 °C goal. Such measures, that are apparently of value for the safeguarding of the planetary condition in the present situation, can also be seen as representing what is to be expected in ethical terms. So far, they remain important assets in the attempts to broaden the scope of social agency within the dynamics of a crisis transformation. All this contributes to the emergence of a globally effective governance of knowledge. As Valentin-Yves Mudimbe notes earlier in another context, African societies, especially vulnerable to the impacts of climate change today, 'have been established as realities for knowledge' (Mudimbe, 1988: XI) that need to be recognised. Hence, African societies also genuinely represent climate change knowledge as recognised and imposed on them by the power of external governance actors and discourse agents. This is what the IPCC is doing through its mission of constant communication of global climate change realities and its assessment of the state of research that serves as the basis for any future climate action that responds to these findings and the IPCC's vision of the 'future as cultural fact' (Appadurai, 2013). The epistemological debate is a major dimension (Mudimbe, 1988) of the processes of sociation and crisis response. As a transnational institution, the IPCC has established rather effective ways of discourse governance of global climate change knowledge, as any dissemination of this knowledge is sampled and thus provided with the symbolic authority of the IPCC seal of approval. As Silke Beck and co-authors point out, thus far the 'IPCC has been hegemonic' (Beck et al., 2014: 84) in determining how to approach and epistemologically govern climate change. As a landmark institution, the IPCC collects data on climate change which it considers relevant to the global condition and bases its reports meant to inform transnational strategies and the actions of national governments on that data. As it has been established as the lead global institution that directs climate change knowledge, defines which knowledge is relevant for global assessment and distributes it to stakeholders, its position as the epistemological centre of the climate change discourse is mostly undisputed. This is not least due to the above-mentioned agenda of an approach devoted to universal equity in the affectedness by climate change and to the neutrality of the reports that do not draw conclusions about responsibility, injustice, etc.

The climate change equity approach favoured by the IPCC addresses climate change as a coherent universal phenomenon of global significance that eventually focuses on calculating average values of global warming. Today, this would be 1.5 to 2 °C, if the goals from the 2015 Paris Agreement were successful. Yet the UN itself sees this as questionable. 'Halfway to the deadline for the 2030 Agenda [...] we are leaving more than half the world behind', UN Secretary-General Guterres criticises in his Foreword to the UN *Sustainable Development Goals Report 2023* (UN, 2023: 2), and continues:

> Progress on more than 50% of targets of the SDGs is weak and insufficient; on 30%, it has stalled or gone into reverse. These include key targets on poverty, hunger and climate. (UN, 2023: 2)

Such focus on global universalism in climate change, supported by the global urgency of the problem, creates the appearance of equity in global climate change affectedness and vulnerability.

Important to note when dealing with the global governance of climate change is that there is a rupture in the IPCC's 'gold-standard' provision of knowledge that concerns structural aspects as much as epistemological ones. The IPCC seeks to integrate that rupture when it focuses on a diversity of geographical regions worldwide. However, its neutrality, which is thought of not only in terms of political non-interference but also in methodological and epistemological terms, misses much information about those regions of the world that are considered peripheral to the global organisation of transnational politics. The Southern African region is among these regions due to a lack of technical resources for data production, influence in global scientific institutions, or geopolitical weight (IPCC, 2014, 2023). One specific problem of the IPCC is that due to its approach to data assessment outlined above, the IPCC in many cases lacks data about particular regions or processes. As its 2012 Special Report titled *Managing the*

Risks of Extreme Events and Disasters to Advance Climate Change Adaptation states:

> There is medium confidence that droughts will intensify in the 21st century in some seasons and areas, due to reduced precipitation and/or increased evapotranspiration. This applies to regions including southern Europe and the Mediterranean region, central Europe, central North America, Central America and Mexico, northeast Brazil, and southern Africa. Definitional issues, lack of observational data, and the inability of models to include all the factors that influence droughts preclude stronger confidence than medium in the projections. (IPCC, 2012: 113–14)

The report goes on to point out that there was only

> low to medium confidence in regional trends in heavy precipitation in Africa due to partial lack of literature and data, and due to lack of consistency in reported patterns in some regions. The AR4 reported an increase in heavy precipitation over southern Africa, but this appears to depend on the region and precipitation index examined. (IPCC, 2012: 143)

That means that findings about the probability of climate change development are produced using insufficient equipment in research infrastructure established in a particular region. This issue is still relevant because to this day

> many funding, knowledge and practice gaps remain for effective implementation, monitoring and evaluation of adaptation (*high confidence*), including, lack of climate literacy at all levels and limited availability of data and information (*medium confidence*); for example for Africa, severe climate data constraints and inequities in research funding and leadership reduce adaptive capacity (*very high confidence*). (IPCC, 2023: 28)

The existing inequalities of access to knowledge assessment produce different qualities in the knowledge about global climate change in the different world regions. The IPCC approach to standardise this knowledge as one coherent epistemological resource thus significantly contributes to the production and emergence of global inequalities as epistemic injustice (Fricker, 2007).

Global inequity in scientific infrastructure is thus decisive for the production of climate epistemology as represented by the IPCC. To meet the standards of knowledge assessment defined by the IPCC it would be necessary first to guarantee a comparable level of data production and knowledge assessment worldwide. Otherwise—and paradoxically—it is the successful implementation of a planetary epistemology achieved through the work of the IPCC that would at the same time reinforce processes of imminent epistemic marginalisation. This is the case with the lack of sufficient data in Southern Africa but also in many other regions in the peripheries of the globalised world that still make up the material level of the otherwise more abstract meta-concept of the planetary. The lack of acknowledged data decreases the reliability of the reports which sometimes leads the IPCC to suggest rather modest scenarios. Its focus on a selection of epistemologically influential sources needs to be questioned. Almost completely excluded from the reports are

> alternative forms of expertise—such as legal reports, which, as a rule, are rarely published in these forums—along with more localized and informal forms of knowledge. (Beck et al., 2001: 83)

As Beck and her co-authors state, 'new guidelines for the integration of "grey literature"' have been introduced with the 2014 AR5 and as 'a response to these shortcomings' (Beck et al., 2001: 83). However, even in this and the following reports the acknowledgment of alternative forms of expertise, especially of local experience and knowledge, is still rather marginal and only rarely consulted.

Conclusion

Paradoxically, it is the twofold ethos of the IPCC that produces its shortcomings, sometimes indeed its failures. The ethos of a universal equity-based approach to climate change creates the appearance of unquestionably equal affectedness by climate change worldwide, which greatly obscures the reality of differences in affectedness. The multiplicity of climate change reality is scarcely represented in the IPCC reports. Its regional organisation and the gradation of confidence in climate development distinction into high, medium and low does not emphasise diversity. Instead, it builds an instrument for measurement that hypothesises global equity instead of diversity. As Beck et al. point out, 'there is no such thing as a one-size-fits-all model for the governance of expertise' (Beck et al., 2001: 84). This becomes especially clear in the IPCC's decision to work only with academic literature and reject any other form of expertise. By doing so, the panel re-establishes a particular form of epistemological hegemony which not only ignores different standards of research in different world regions, but above all ignores existing expertise, experiences and knowledge in many regions worldwide. Thus, by pursuing its intent to act as the leading institution for the provision of reliable climate change knowledge on which governments can rely in designing climate governance measures, the IPCC is creating a sort of postcolonial epistemological climate change regime.

References

Appadurai, A. (2013). The future as cultural fact: Essays on the global condition. London & New York: Verso

Adorno, Th.W. (1997). *Minima Moralia*. Frankfurt a.M.: Suhrkamp.

Balbi, S., Bagstad, K. J., Magrach, A., Sanz, M. J., Aguilar-Amuchastegui, N., Giupponi, C., & Villa, F. (2022). The global environmental agenda urgently needs a semantic web of knowledge. *Environmental Evidence*, 5/2022, 1–6. https://doi.org/10.1186/s13750-022-00258-y

Beck, S., Borie, M., Esguerra, A., & Chilvers, J. (2001). Towards a reflexive turn in the governance of global environmental expertise. The cases of the IPCC and the IPBES. *GAIA*, 23/2, 80–87

Beck, S. (2009). *Das Klimaexperiment und der IPCC. Schnittstellen zwischen Wissenschaft und Politik in den internationalen Beziehungen*. Marburg: Metropolis-Verlag

Beck, S. (2011). Moving beyond the linear model of expertise? IPCC and the test of adaptation. *Regional Environmental Change*, 11/2011, 297–306

Beck, S. et al. (2014). *Towards a reflexive turn in the governance of global environmental expertise: The case of the IPCC and the IPBES. Gaia-Ecological perspectives for science and society*. 32(2), 80–87.

Beck, S., & Mahony, M. (2018). The IPCC and the new map of science and politics. *WIREs Climate Change*, 1–16. https://doi.org/10.1002/wcc.547

Beck, U. (2010). Climate for change, or how to create a green modernity? *Theory, Culture & Society*, 27(2–3), 254–266

Blau, J. (2017). *The Paris agreement: Climate change, solidarity, and human rights*. Cham: Palgrave Macmillan. https://doi.org/10.1093/sf/soy055

Boykoff, M., & Pearman, O. (2019, November 22). Now or never: How media coverage of the IPCC special report on 1.5 °C shaped climate-action deadlines. *One Earth*, 1, 285–288. https://doi.org/10.1016/j.oneear.2019.10.026

Chakrabarty, D. (2009). The climate of history: Four theses. *Critical Inquiry*, 35(2), 197–222

Chakrabarty, D. (2012). Postcolonial studies and the challenge of climate change. New Literary History, 43(1), 1–18

Chakrabarty, D. (2016). Whose Anthropocene? A response. *RCC Perspectives*, 2/2016, 101–114

Chakrabarty, D. (2018). Planetary crises and the difficulty of being modern. *Millennium: Journal of International Studies*, 46(3), 259–282. https://doi.org/10.1177/0305829818771277

Collyer, F. M. (2018). Global patterns in the publishing of academic knowledge: Global North, global South. *Current Sociology*, Vol. 66, 1/2018, 56–73

Dasgupta, S. (2023). Inequality and growth impacts of climate change—insights from South Africa. *Environmental Research Letters*, 18/2023, 1–21. http://dx.doi.org/10.1088/1748-9326/ad0448

Escobar, A. (2020). Thinking-feeling with the earth: Territorial struggles and the ontological dimension of the epistemologies of the South. In B. de Sousa Santos, & M. P. Meneses (Eds.), *Knowledges born in the struggle: Constructing the epistemologies of the global South* (pp. 41–57). New York & London: Routledge. http://dx.doi.org/10.11156/aibr.110102e

Fleck, L. (1980). *Entstehung und Entwicklung einer wissenschaftlichen Tatsache. Einführung in die Lehre von Denkstil und Denkkollektiv*. Frankfurt a.M.: Suhrkamp. (Original work published 1935)

Fricker, M. (2007). *Epistemic injustice: Power and the ethics of knowing*. Oxford: Oxford University Press

Galaz, V., Biermann, F., Crona, B., Loorbach, D., Folke, C., Ollson, P., Nillson, M., Allouche, J., Persson, A., & Reischl, G. (2012). 'Planetary boundaries'—exploring the challenges for global environmental governance. *Current opinion in environmental sustainability*, 4/2012, 80–87

Grundmann, R., & Stehr, N. (2010). Climate change: What role for sociology? A response to constance Lever-Tracy. *Current Sociology*, 58(6), 897–910

Guterres, A. (2020, December 2). *Secretary-General: 'Making peace with nature is the defining task of the 21st century'. Landmark speech on the state of the planet at Columbia University in New York*. Retrieved January 23, 2025, from https://www.unfccc.int/news/un-secretary-general-making-peace-with-nature-is-the-defining-task-of-the-21st-century

Guterres, A. (2023, July 27). *Secretary-General's opening remarks at press conference on climate, United Nations*. Retrieved January 20, 2025, from https://www.un.org/sg/en/content/sg/speeches/2023-07-27/secretary-generals-opening-remarks-press-conference-climate

Hardt, M., & Negri, A. (2002). *Empire. Die neue Weltordnung*. Frankfurt/New York: Campus

Hartnett, R. (2021). Climate imperialism: Ecocriticism, postcolonialism, and global climate change. *eTropic: electronic journal of studies in the tropics*, 2021/2, Special Issue: Tropical imaginaries and climate crisis, 138–155. https://doi.org/10.25120/etropic.20.2.2021.3809

Hartz, F. (2022). Leaking the IPCC: A question of responsibility? *WIREs Climate Change*, 14, e814, 1–8. https://doi.org/10.1002/wcc.814

Hulme, M. (2009). *Why we disagree about climate change: Understanding controversy, inaction and opportunity*. Cambridge, UK, & New York: Cambridge University Press

Hulme, M., & Mahony, M. (2010, June 18). Climate change: What do we know about the IPCC? *Progress in Physical Geography Online*

IPCC (2007). *Climate change 2007: Impacts, adaptation and vulnerability*. Contribution of working group II to the fourth assessment report of the intergovernmental panel on climate change, M. L. Parry et al. (Eds.). Cambridge: Cambridge University Press

IPCC (2012). *Special report on managing the risks of extreme events and disasters to advance climate change adaptation*. Retrieved February 8, 2024, from https://www.ipcc.ch/report/managing-the-risks-of-extreme-events-and-disasters-to-advance-climate-change-adaptation/

IPCC (2014). *Climate change 2014: Impacts, adaptation, and vulnerability*. part A: Global and sectoral aspects. Contribution of working group II to the fifth assessment report of the intergovernmental panel on climate change, C.B., V.R. Barros et al. (Eds.). Cambridge: Cambridge University Press

IPCC (2018). *Special report global warming 1.5 °C*. Ch. 5: Sustainable development, poverty eradication and reducing inequalities. Retrieved February 8, 2024, from https://www.ipcc.ch/sr15/chapter/chapter-5/

IPCC (2019). *IPCC special report on the ocean and cryosphere in a changing climate*, H.-O. Pörtner et al. (Eds.). Cambridge & New York, NY: Cambridge University Press

IPCC (2022a). Summary for policymakers. In *Climate change 2022: Mitigation of climate change*. Contribution of working group III to the sixth assessment report of the intergovernmental panel on climate change, P.R. Shukla et al. (Eds.). Cambridge & New York, NY: Cambridge University Press. https://dx.doi.org/10.1017/9781009157926

IPCC (2022b). *AR6: Climate change 2022: Mitigation of climate change*. Working group III contribution to the sixth assessment report of the intergovernmental panel on climate change. Retrieved February 8, 2024, from https://www.ipcc.ch/report/ar6/wg3/

IPCC (2023). *Climate change 2023: Synthesis report*. Contribution of working groups I, II and III to the sixth assessment report of the intergovernmental panel on climate change [Core Writing Team, H. Lee and J. Romero (Eds.)]. Geneva, Switzerland: IPCC

IPCC. (n.d.). *About the IPCC*. Retrieved February 8, 2024, from https://www.ipcc.ch/about/

Islam, N., & Winkel, J. (2017, October). *Climate change and social inequality*, DESA Working Paper No. 152 ST/ESA/2017/DWP/152, 1–30

Jacobson, S. G. (Ed.) (2018). *Climate justice and the economy: Social mobilization, knowledge and the political*. London & New York: Routledge

Jamison, A. (2010). Climate change knowledge and social movement theory. *WIREs Climate Change*, 1, 811–823

Jasanoff, S. (2010): *A new climate for society, theory, culture & society*, Vol. 27, 2–3/2010, 233–253

Jazeel, T. (2011). Spatializing difference beyond cosmopolitanism: Rethinking planetary futures. *Theory, Culture & Society*, Vol. 28, 5/2011, 75–97

Kashdan, A., & Ostanek, K. (2022, February 18) *What to know about the UN's major upcoming climate change reports*. United Nation foundation. Retrieved February 8, 2024, from https://unfoundation.org/blog/post/what-to.know-about-major-upcoming-un-climate-change-reports/

Keller, S. (2010). Scientization: putting global climate change on the scientific agenda and the role of the IPCC. *Poiesis und Praxis. International Journal of Ethics of Science and Technology Assessment*, 7/2010, 197–209

Koch, M., Buchs, M., & Lee, J. (2023). Towards a new generation of social policy: Commonalities between sustainable welfare and the IPCC. *Social policies*, 1/2023, 27–42. http://dx.doi.org/10.7389/107137

Livingston, J. E., & Rummukainen, M. (2020) Taking science by surprise: The knowledge politics of the IPCC special report on 1.5 degrees. *Environmental Science and Policy*, 112, 2020, 10–16. https://doi.org/10.1016/j.envsci.2020.05.020

Livingston, J. E., & Rummukainen, M. (2023). Making policy-relevant knowledge in the IPCC special report on 1.5 degrees: An analysis of reviewer comments. *Environmental Science and Policy*, 147/2023, 305–314. https://doi.org/10.1016/j.envsci.2023.06.001

Markkanen, S., & Anger-Kraavi, A. (2019). Social impacts of climate change mitigation policies and their implications for inequality. *Climate Policy*, 7/2019, 827–844. http://dx.doi.org/10.1080/14693062.2019.1596873

Mbembe, A. (2017). *Critique of black reason*. Durham & London: Duke University Press. https://doi.org/10.1215/9780822373230

Mignolo, W. (2011). *The darker side of Western modernity: Global futures, decolonial options*. Durham & London: Duke University Press

Miller, C. A. (2004). Climate science and the making of a global political order. In S. Jasanoff (Ed.), *States of knowledge: The co-production of science and social order* (pp. 46–66). London & New York: Routledge

Mudimbe, V. Y. (1988). *The invention of Africa: Gnosis, philosophy, and the order of knowledge*. Bloomington & Indianapolis: Indiana University Press

Robinson, S. A. (2020). Climate change adaptation in SIDS: A systematic review of the literature pre and post the IPCC fifth assessment report. *WIREs Climate Change*, 2020/11, e653, 1–21. https://doi.org/10.1002/wcc.653

Sanford, M., Painter, J., Yasseri, T., & Lorimer, J. (2021). Controversy around climate change reports: a case study of Twitter responses to the 2019 IPCC report on land. *Climatic Change*, 2021/167, 59, 1–25. https://doi.org/10.1007/s10584-021-03182-1

Selchow, S. (2014, May 15). *Interview with Ulrich Beck and Bruno Latour*. Frankfurter Allgemeine Zeitung

Shoshitaishvili, B. (2022). Is our planet doubly alive? Gaia, globalization, and the Anthropocene's planetary superorganisms. *The Anthropocene Review*, 10(2), 434–454. https://doi.org/10.1177/20530196221087789

Sleeman, J., Finin, T., & Halem, M. (2018). Ontology-grounded topic modeling for climate science research. In E. Demidova et al. (Eds.), *Emerging topics in semantic technologies* (pp. 1–10). Berlin: AKA Verlag. https://doi.org/10.48550/arXiv.1807.10965

Sousa Santos, B. de (2016). Epistemologies of the South and the future. *From the European South*, 1/2016, 17–29

Sousa Santos, B. de (2018a). *Epistemologien des Südens. Gegen die Hegemonie des westlichen Denkens*. Münster: Unrast

Sousa Santos, B. de (2018b). *The end of the cognitive empire: The coming of age of epistemologies of the South*. Durham & London: Duke University Press. https://doi.org/10.1215/9781478002000

Taylor, C. (2004). *Modern social imaginaries*. Durham, NC: Duke University Press

Tol, R. S. J. (2011). Regulating knowledge monopolies: The case of the IPCC. *Climatic Change*, 108, 827–839

Vanderheiden, S. (2008). *Atmospheric justice: A political theory of climate change*. Oxford & New York: Oxford University Press

Yearley, S. (2009). Sociology and climate change after Kyoto: What roles for social science in understanding climate change? *Current Sociology*, 57(3), 389–405

FSC
www.fsc.org
MIX
Papier aus verantwortungsvollen Quellen
Paper from responsible sources
FSC® C105338

If you have any concerns about our products,
you can contact us on
ProductSafety@springernature.com

In case Publisher is established outside the EU,
the EU authorized representative is:
Springer Nature Customer Service Center GmbH
Europaplatz 3, 69115 Heidelberg, Germany

Printed by Libri Plureos GmbH
in Hamburg, Germany